MODIFYING MAN:

Implications and Ethics

Edited by Craig W. Ellison

University Press
of America™

ISBN: 0-8191-0302-0

TABLE OF CONTENTS

iv

PREFACE

This book began in March 1973 with the idea for a conference that would draw evangelical Christian scholars from a variety of disciplines together to systematically consider the complex issues being raised by the advent of human engineering possiblilities. It had been my observation that all too frequently those from theologically conservative circles did not address the important issues of their time until after the fact. Then, many of the responses have tended to be reactive and negative in tone, rather than constructive and positive.

As plans for the conference developed it became apparent that a number of Christian organizations felt that such a gathering would be important, as evidenced by the substantial number of organizational co-sponsors which contributed financially toward the project (see list of co-sponsors in Section VIII). It also was obvious that much of the work in these fields had been done by those who were not committed to an evangelical Christian position, but who had keenly begun to raise issues of common concern. As a result, it was felt that dialogue between non-evangelical experts in the respective fields and an interdisciplinary variety of evangelical scholars would be helpful. Consequently all of the responses to the major papers are by evangelicals, while four of the six major addresses are from noted authorities who have consistently raised important ethical questions while not sharing exactly the same philosophical and theological beginning point. In addition, there is considerable theological and philosophical variety represented by the evangelical respondents. The result was and is, I believe, a valuable and interesting dialogue that was begun.

It is important to realize that this conference and book represent only a beginning point. It was the first such conference. To date no similar or follow-up conference on these issues has been held, though such is sorely needed. The preliminary conclusions of the conference small-groups and the conference commission were surprisingly positive. There was a sense of moral and spiritual responsibility that would not allow a quick rejection of human engineering research and application. At the same time a number of important constraints and cautions were sounded. The result, I think, was a healthy balance of compassion and wisdom. There was conscious recognition that such a position means a continuing process of evaluation and tension. It is not the easy road of complete rejection or total embrace.

vii

It is my hope that the conference and this book will stimulate much constructive dialogue within and between the scientific, educational and religious communities.

A number of persons were significant in various stages of this project. I am especially grateful for the enthusiastic encouragement and active support that Dr. Lewis Bird contributed throughout the planning of the conference and the preparation of the book. His energy, optimism and insight kept the project from floundering at critical points. Also the personal involvement and competent leadership of Bill Sisterson and Elving Anderson were vital throughout the planning and implementation of the conference.

The confidence and support of the co-sponsoring organizations were vital for the implementation of the conference idea. I am grateful to each of them for responding to my plea concerning the need for such a conference by providing both income and leaders to help plan the conference. The Christian Medical Society kindly provided space for the Planning Council to periodically meet. Mrs. Rachel Buick was an extremely gracious and capable secretary-hostess for those meetings.

The conference could not have been held and the book would not have been published without timely and generous financial assistance as well from Dr. Robert Scheidt, Mr. Paul Johnson, and Mr. Michael Woodruff. Each of these individuals are from widely divergent professions, but each felt that the projects were significant enough to warrant their support.

I am indebted to a large number of persons for the preparation of the book. Cindie Ryan ably persevered through the typing and retyping of the majority of the manuscript. Dorothy Hawk and Anita Hillje provided time and skill to help complete the book, as well as many hours in proofreading. I am grateful also to Dr. J. Randall Springer for his exceptionally competent assistance in copy editing the major portion of the manuscript. Cherie Groeneveld also contributed many hours of her skills in proofreading. I am grateful as well to Jenny Whitmore, Betty Spears, and Don Evans for their help in proofing. Stewart Ensign and Linda Timms read and provided helpful feedback for portions of the text.

Finally, I am grateful to Westmont College administrators Dr. George Brushaber and Dr. Ernest Ettlich, and to Dr. Bruce Stockin, Chairman of the Psychology Department, for their strong encouragement and support throughout the many months of conference planning and book preparation.

My deepest appreciation is reserved for my wife, Sharon, for her grace and understanding throughout the many hours of planning, separation and writing required by the project.

PART I
AN OVERVIEW OF THE ISSUES

THE ETHICS OF HUMAN ENGINEERING

Craig W. Ellison
Westmont College

A revolution is underway. Its potential effects have significant implications for individuals, families, and whole
societies. Its ultimate impact could be more pervasive and dramatic than that of the Copernican, Industrial and Darwinian
revolutions. Like these, it is the product of scientific investigation and technological application.

For the first time in human history it is becoming possible
to systematically and precisely modify human genetic and behavioral characteristics. The combined results of current and
future research in genetics, brain physiology, pharmacology, and
psychology present contemporary civilization with the prospects
of fundamental changes in human values, functioning and relationships.

The implications of the Human Engineering Revolution are
awesome. The search for an ethical framework to guide research
and application of new knowledge is frequently frustrating
and confusing. One is reminded of the saying, "There is no
subject, however complex, which--if studied with patience and
intelligence--will not become more complex."

Throughout this book human engineering will be used to mean
*systematic attempts to change human functioning through the in-
vestigation and application of scientific knowledge.* It refers
to the modification of human beings through genetic, physiological, or psychological intervention. The term is not to be
confused with engineering psychology, which is basically the
study of man-machine interaction. It is useful as a general
term for describing the biological and psychological modification of persons.

Craig W. Ellison is Associate Professor of Psychology, Westmont
College, Santa Barbara, California. He was the organizer and
Director of the International Conference on Human Engineering and
the Future of Man, upon which this book is based. Dr. Ellison
is a social-developmental psychologist. He has edited two previous books: THE URBAN MISSION (1974) and SELF ESTEEM (1976).
He has also written a number of articles for professional and
popular periodicals. Dr. Ellison is the newly appointed editor
of the "Christian perspectives on counseling and the behavioral
sciences" series, published by Harper & Row and the Christian
Association for Psychological Studies.

3

Before we delve more specifically into some of the complex ethical, theological and value issues raised by human engineering, we will briefly describe the technology. It should be noted that most of the procedures described have a host of complex ethical and moral issues associated with them, either with regard to the development of the techniques or their general implementation.

Selection or engineering of human characteristics can be instituted at four points: (1) preconception; (2) conception; (3) prenatal; (4) postnatal. Selection refers to genetically oriented choice procedures which do not involve direct manipulation of the genetic material. Engineering refers to direct manipulation of the genetic material, the environment, or the person. Control may be exercised through either procedure, though the precision is greater with engineering.

GENETIC CONTROL

In the preconception period, control of the quantity of persons may be exercised through *contraception* or *sterilization*. Indirectly, these procedures also affect the appearance of defective and non-defective genetic characteristics. When used in conjunction with *genetic screening* either of these procedures may serve as quality control techniques. That is, if genetic screening suggests that there is a high probability that the couple would produce a seriously defective baby, a choice can be made about reproduction. Such selective reproduction ultimately will affect the human gene pool and the occurrence of genetic defects. However, it will probably never result in the removal of all defects because of mutations and incomplete knowledge about deleterious genes due to their recessive character.

It is estimated that up to five million couples are in need of genetic counseling. Only about 6% of that total are currently involved each year. More than 2000 genetically distinct defects have been identified. Approximately 250,000 defective births occur annually in the United States; about 20% are due to known genetic causes. One percent of all children born have an abnormal number of chromosomes; fifteen percent of newborn infants have hereditary disorders. The annual cost of institutionalization for Downs syndrome alone is $1.7 billion.[1] Today it's the couple's decision to risk having defective children. Tomorrow, it may well be the government's. Over-population, famine, and cost of care may bring about a change in the traditional right to procreate.

4

The problem with the genetic screening technique is that these couples may end up childless if they are prevented from or choose not to take the risk. This is where some of the other techniques begin to become attractive. Alternative means of reproduction or refined techniques of genetic engineering which help avoid the high social cost of the highly dependent defective, while providing the possibilities of parenthood begin to seem much less repulsive to many.

Among the selection and engineering techniques currently being developed for use during the conception period are: (1) artificial insemination by husband or donor; (2) *in vitro* fertilization; (3) cloning.

Artificial insemination involves the mechanical fertilization of the egg cell by the insertion of semen into the recipient's uterus from either a husband or another donor. A donor may be used in cases of husband infertility, or as a eugenic technique. (Eugenics is the use of various procedures with the purpose of improving the human gene pool). Artificial insemination is, of course, an imprecise selection procedure.

In vitro fertilization is a technique in which an egg is removed from the ovary, fertilized in the laboratory, and returned to the uterus of a hormonally prepared female. This procedure increases the technician's control by allowing selection of egg and (potentially) sperm. The "mother" may or may not be the original donor of the egg. One British scientist has claimed success in the fertilization, gestation and birth of three separate children using this technique, although he has not produced documented evidence. Two other British scientists have documented conception using this method, though the fetus died after ten weeks because of a tubal pregnancy. It has been estimated that from .5-1% of all women might be helped to have a child of their own by this technique.[2]

Cloning involves removing an egg cell, removing its nucleus, substituting an adult body cell nucleus, culturing the fetus in a laboratory medium and implanting the renucleated egg in a prepared uterus. Cloning is asexual reproduction. If the donated cell were from the female egg donor, the offspring would be genetically identical with the donor of the body cell nucleus. It is probable that a cloned mammal will be born within a few years. The major difficulty remaining is introducing the donor nucleus into the enucleated egg.

The major techniques being developed for use during the prenatal period are: (1) amniocentesis; (2) gene transfer or surgery; and (3) ectogenesis. Of these procedures amniocentesis

5

is currently being applied and ectogenesis is probably the
farthest from realization.

   Amniocentesis is the procedure of withdrawing a small amount
of amniotic fluid from the uterus. This is done 12-16 weeks
after conception. The fluid contains cells derived mostly from
the skin and respiratory tract of the fetus. The cells are
then removed from the fluid and cultured. Examination of the
cultured cells is able to identify chromosomal abnormalities
and over 140 biochemical deficiencies. It is estimated that
over 500 metabolic genetic deficiencies will be determined with
this method by 1980.[3] Amniocentesis by itself is only a diagnostic device. If the fetus is defective, a choice must then
be made between several options: (1) the fetus may be allowed
to develop and be born genetically defective; (2) the fetus may
be aborted; (3) at some point in the future, genetic surgery
may be performed; (4) other medical treatments such as fetal
blood transfusions or fetal surgery may be performed. At least
the first three options raise a number of ethical issues.

   Genetic surgery is the purest form of genetic engineering.
It involves changing the genome by means of special enzymes and
bacterial plasmids. One proposal would use restriction enzymes
which are able to cut certain base sequences of the DNA molecules into fragments, some of which are whole genes. The ends
of these fragments are "sticky," and would be able to recombine
with other DNA molecules to form a hybrid molecule that would
not be defective. This may also make it possible to produce
bacteria that will synthesize medically valuable substances
such as insulin, human antibodies, and viral proteins for vaccine production. Eventually, direct manipulation of human
genes may be able to free people from genetic defects. Another
enzyme that is being used is reverse transcriptase. This re-
verses the normal gene action (DNA→RNA) so that a specific unit
of DNA for a particular cell substance can be formed and isolated (RNA→DNA).[4] Although it is likely that direct manipulation of the genes will occur eventually, there are many technical problems that will have to be overcome before the technique
is precise enough to be useful and not detrimental. In addition, serious questions have recently been raised by the scientific community about the safety of doing research with recombinant DNA, though the development of strict standards for
physical and biological containment of potentially dangerous
recombinations seems to have shifted the argument to long-term
fears about the genetic engineering of human beings.[5]

   Ectogenesis is fertilization and gestation of a fetus in an
artificial placenta or fetal incubator. A baby "born" in this
way would be the true "test tube baby," although the stainless

steel apparatus being developed as an artificial placenta is
larger and considerably more complex than a test tube. Spon-
taneously aborted fetuses have been kept alive for up to 48
hours in the fetal incubator. The major problem is disposal
of wastes. Lambs have been kept alive for up to 55 hours, and
mouse and rat embryos have been kept alive through the heart
beating stage.[6] Although the cost of developing ectogenesis
would be extensive and it would probably not be used as a
normal, mass-applied technique, it could be used to aid in the
development of more efficient fetal immunization and genetic
engineering procedures by making the fetus more accessible.

A variety of postnatal, genetically oriented techniques are
also being developed. These procedures are collectively referred to as euphenics. They include such approaches as
direct injection of a missing enzyme, implantation of a missing
enzyme under the skin so that small amounts leak into the blood-
stream, and actual gene manipulation (a future possibility).
In addition, nutritional control and organ transplants may be
used.

PSYCHOLOGICAL CONTROL

Proponents of eugenics actually wish to produce an ideal
phenotype rather than the perfect genotype. This raises serious difficulties because most normal human characteristics are
polygenic in character. Also, mental and social traits are
dependent upon environmental influences. For this reason the
eminent geneticist, Sir Francis Crick, has stated that he feels
"humans are less likely to improve by newer genetic technologies
than by manipulation of the environment or enhancement of educational technologies."

There are eight psychologically-related techniques by which
human beings can have their attitudes, feelings, beliefs,
values, states of consciousness or behavior altered. These
techniques may be divided into two categories: (1) biopsychological intervention; (2) environmental manipulation.

Biopsychological intervention is accomplished through the
use of psychoactive psychosurgery, drugs, or electrical stimulation of the brain. These procedures involve some form of
direct physical intervention, primarily focused on the brain.
There are resulting psychological and behavioral changes. These
techniques promise control of behavior from within, by changing
the internal state of the organiam. They bypass normal sensory
and information processing mechanisms.

Environmental manipulation includes behavior conditioning,
information control, "brainwashing," psychotherapy, and

7

biofeedback. These techniques involve feedback from the external or internal environment. The organism processes the feedback, although not always consciously, and the information received is capable of altering the subsequent responses of the organism.

Of all the psychological control techniques, psychoactive drugs and information control seem to have the greatest potential for mass use. At this point behavior conditioning and biofeedback seem to be the most precise and predictive in influence. For the most part these psychological techniques are employed postnatally, although the possibility of using them to alter the prenatal environment is a real one.

BIOPSYCHOLOGICAL INTERVENTION

Psychosurgery refers to the destruction of a portion of the brain in order to modify or eliminate undesirable behaviors or states of mind. Usually, organically healthy portions of the brain are destroyed through use of a cutting instrument or electrolytic lesion. These may be frontal or temporal areas, or in the limbic system. Psychosurgery seems to influence subsequent behavior by affecting the brain's overall capacity to respond emotionally.[7] The person seems to be generally subdued rather than having a specific behavior pattern affected. The most predominant use of psychosurgery has been to control pathological aggression, or violent and uncontrollable behavior. The target population seems limited primarily to: (1) children who demonstrate unmanageable hyperactivity and violent behavior, including many who are also severely retarded or epileptic; (2) epileptic patients in whom violent and aggressive behavior appears either as part of the seizure manifestation or related psychopathology; and (3) institutionalized persons who are violent prisoners or emotionally ill. Due to the advent of modern psychoactive drugs, psychosurgery seems to have declined from over 70,000 procedures in the late 1940s-mid 1950s to about 500=700 per year currently in the United States. These are performed by about 50 surgeons. In comparison about 250 million Americans have taken neuroloptic drugs since 1950.

As many as 37 million Americans use sedatives such as barbituates legally and illegally. They spend $2.5 billion annually on psychoactive drug prescriptions, and at least $2 billion more on illegal drugs.[8] Psychoactive drugs have the highest rate of entry of any legitimate drugs. The National Institute of Mental Health lists over 1000. The list is expanding rapidly. New research findings are producing more specificity and direct action with fewer side effects.

8

The stimuli behind this tremendous explosion of drug use are multiple. It is partially due to the discovery that tranquilizers virtually eliminated the need for psychosurgery or strait jackets, substantially reduced the use of electric shock therapy, and made it possible to place patients in the community at much lower cost than long-term hospitalization. In addition, the "happiness ethic"[9] of Americans has promoted widespread drug use. Drugs are used to cope with more or less normal life stresses, depression and boredom in wholesale fashion. Many critics argue that doctors, who are largely dependent upon pharmaceutical house ads for their information, have encouraged over-use. It has been estimated that 90% of psychoactive drugs have been prescribed by general practioners and internists, not psychiatrists or neurologists.

Psychoactive drugs are chemical agents which affect the psychological state or behavior of the user. Their major purposes are as therapeutic agents, for recreational use, and as performance enhancers.[10] Theraupeutic drugs include: (1) the antipsychotic drugs or major tranquilizers, such as the phenothiazines. The prototype of the phenothiazines is chlorpromazine (Thorazine). The antipsychotic drugs produce sedative, hypnotic and mood elevating effects. They are used to treat major mental illnesses. They do not cure the disorders but suppress the symptoms, resulting in reduced hospitalization time; (2) the anti-depressant drugs. Among these are the psychomotor stimulants, such as the amphetamines. Also Ritalin and Preludin, widely used in relation to minimal brain disfunction and hyperactivity among school children belong to this class of drugs. The tricyclic derivatives of imipranine are another sub-group of anti-depressants. These are the most widely used of the anti-depressant drugs, and are commonly known as Tofranil, Elavil and Pertefrane; (3) anti-anxiety drugs, or minor tranquilizers, are used to treat temporary episodes of neurotic symptoms typically related to situational stress. Long-term usage tends to produce psychic dependence and abuse. These are the most widely prescribed drugs in the United States. They include the barbituates, which are the least effective and highest risk drug in the group, and the heavily used diazepoxides, commonly known as Librium, Valium and Serox.

In addition to drugs used for therapeutic purposes, increasing numbers of people are using drugs such as alcohol, marijuana, psychedelics, and cocaine for recreational or personal pleasure. It has been estimated, for example, that over 13 million Americans smoke marijuana regularly, and over 89 million spend $25 billion annually for alcoholic beverages.[11] These drugs are frequently taken in attempts to relieve depression

9

and to experience a high that lifts people out of feelings of meaninglessness and boredom.

Also, a new class of drugs to enhance performance and capabilities seem to be developing. Some feel that developments in this area will represent the most significant uses of drugs in the future. Currently, caffeine and the amphetamines, used to relieve fatigue temporarily, are among the few available. Research by such psychologists as James McConnell, in which it appears that learning can be transmitted chemically (at least among flatworms fed a diet of fellow trained flatworms), suggests the possibility of drugs that will improve learning and memory. It is equally possible that other functions can be equally improved by the simple ingestion of appropriate drugs.

The third form of biopsychological intervention currently being developed is *electrical stimulation of the brain* (ESB). This technique involves implantation of electrodes into specific regions of the brain, and the discharge of specified electrical current through the electrodes. The intensity, frequency, and duration of stimulation can be varied and controlled with or without the subject's knowledge. ESB has not yet been accepted as a standard therapeutic technique, but is likely to become more accepted because of its flexibility and non-destructive effects.

Most ESB investigation has focused on the septal region, and the hypothalamus and thalamus in particular. The production or inhibition of rage, sleep, motor, functions, calming, and sexual activity have all been demonstrated through ESB. After stimulation the mood of schizophrenics seems to improve for several days. It has been shown to help relieve anxiety, depression and epileptic seizures. Some research involving epileptics suggests that ESB can activate human pleasure centers so that self-stimulation becomes highly reinforcing.[12] It is important to note, however, that electrodes placed in a specific part of the brain do not always produce a particular behavior. In a large percentage of cases, animals don't show any specific behavior in response to ESB. Also, even in animals, which have more stereotypic behavior than humans, (1) stimulation of the same brain region produces variable results both within and between organisms, and (2) electrodes placed at different sites in the same organism may produce the same behaviors. Exact prediction of responses to specific ESB is clearly not possible at present, although once individual brain circuitry is established prediction improves.[13]

The second class of psychological control techniques include information control, "brainwashing," conditioning, biofeedback, and psychotheraphy. We will focus on the first four. Psychotherapy in its different forms is perhaps the most well known, and, in the opinion of some, the least effective modification procedure.

Information control is the selective provision of information. Through information human beings form and change attitudes and values, establish emotional responses, and direct their behaviors. Information comes both through self-observation and through a wide variety of external sources. It may be obtained directly from the source of information, as in the case of face-to-face interaction, or indirectly, through friends, television, radio, newspapers, etc. Information not only provides a basis upon which to act, but it also shapes philosophical orientations and self-concepts. It does the latter by providing both ideals and evaluational feedback. One's self-concept, in turn, strongly influences behavior. The two most powerful contemporary forms of information control are the educational system, and mass media, particularly television. Both technologies provide implicit and explicit ideals and norms which are either directly or vicariously reinforced.

Biofeedback, a variant of information control, is primarily used to relieve stress. It involves the monitoring and control of various autonomic nervous system responses by means of feedback (auditory or visual) provided by a machine to which the subject is connected by means of electrodes. It has been discovered that brain waves, heart rate, respiration rate, and other internal processes previously thought to be outside of conscious control can be controlled through training and use of the biofeedback apparatus. Production of alpha waves has resulted in the relief of stress, as deep relaxation is associated with that level of brain wave activity. The procedure is one of conscious self-control, with the aid of the machine feedback.

Conditioning is probably the most recognized psychological control technique, as a result of the furor generated by B. F. Skinner's BEYOND FREEDOM AND DIGNITY. Actually there are two forms of behavior conditioning: respondent and operant. Respondent conditioning involves the temporal pairing of a previously neutral stimulus with an unconditioned stimulus that is biologically linked with a natural or unconditioned response. Through this association the neutral stimulus becomes capable of eliciting the unconditioned response. The prototype of respondent, or classical conditioning, is the Pavlovian dog which salivated at the sound of the bell. Operant conditioning is the positive or negative reinforcement of a given behavior, or operant response. In general, positive reinforcement increases the emission probability and rate, negative reinforcement decreases it. Behaviors can be progressively refined or shaped through the selective application of reinforcement to successive approximations of the desired behavior. Operant conditioning has become widely applied through the development of token economies. It is used in hospitals, schools, smoking clinics, diet watchers' clubs, and prisons. It has been applied to a wide variety of dysfunctional or unwanted behaviors, with considerable success.

"Brainwashing" is a complex mixture of information control, conditioning, and coercion. It involves total control of the environment by the modification agents. It does not have to involve physical torture, psychoactive drugs, or invasion of the brain through surgery or stimulation to be effective. Brainwashing was first used in its modern form on American prisoners of war by the Communist Chinese during the Korean conflict. Because it necessitates total environmental control it is most likely to be applied by dictatorial governments. It is perhaps least likely to be used in a free America. However, the advent of religious "deprogramming" which has developed as a means of trying to modify the behavior/attitudes of young people involved in a variety of contemporary cults is highly similar to classic brainwashing. It has become increasingly employed by parents trying to win their young adult children back to their own religious beliefs.

Common to both classic brainwashing and deprogramming are attempts to break down trust, the removal of personal emotional supports, physical exhaustion, implied or overt threats, positive reinforcement for compliance, repetition, and group pressures to conform. These are molded together in an attempt to break down the emotional and intellectual defenses of the individual so that the orientation desired by the controllers is adopted. There is some question as to the permanence of such "changes," however. Outside of the controlled environment, a high proportion of individuals seem to revert back to their original orientation.

The subject of human engineering typically creates polarized and passionate responses. Dialogue easily degenerates into accusation and shouting. Both sides of the issue implicitly understand the high stakes.

For many scientists the issue centers around the *freedom of inquiry*. Until recent years this right to know was relatively unchallenged. It has been the closest thing to an absolute that most scientists might defend. Freedom of inquiry is based on two underlying premises: (1) that inquiry itself is morally neutral.[14] Although this particular premise has been under attack since the development and use of the atom bomb, there are many scientists who hold to a distinction between pure and applied research. Their argument is that the act of "pure" scientific investigation is objective and value-free. Knowing is morally neutral within this paradigm. Therefore, scientists do not bear moral responsibility for the ways in which their findings are applied; (2) that progress is the highest good. The argument here is partially theoretical and partially historical. On theoretical grounds, belief in evolutionary development led many to adopt an unbridled optimism about human progress. The occurrence of two world wars and many other disillusioning events in this century has introduced caution into the optimism. The argument has become that progress can only be guaranteed if the human race takes more active control of the evolutionary process and shapes its future. On historical grounds, the scientist is able to point to the considerable material improvements that have been made through the discoveries and application of science. These improvements have allowed people to live healthier, longer and more comfortable lives.

The continuing right of scientists to inquire into all aspects of human existence is then seen as essential to both the progress and the very survival of the human race. There is both an implicit and explicit obligation to future generations which is considered moral in nature. This obligation is then used as a justification for research procedures.

It is important to point out that the motivation behind human engineering research is, for the most part, highly commendable and noble. That is, the desire of the scientists involved is to: (1) alleviate suffering by the correction of genetic or behavioral defects; (2) therapeutically control and rehabilitate those who are societally dangerous; (3) improve the overall functioning and future potential of the human race. It may be argued that the techniques employed are not appropriate, or that it is not the human right to do such things,

13

but few would argue with the goals themselves. The typical scientist is not out to gain personal power over the human race. His brief is simply to help the human condition to the best of his ability.

On the other side are those who argue that the right to know is not an absolute right. Rather, the very act of investigation frequently raises moral issues. According to Paul Ramsey, in THE PATIENT AS A PERSON, the ethical quality of an experiment is not determined by its results or application, but resides in the very design of the experiment. Hans Jonas points out that:

> .... not only have the boundaries between theory and practice become blurred, but....the two are now fused in the very heart of science itself, so that the ancient alibi of pure theory and with it the moral immunity it provided no longer hold.[15]

(Parenthetically, it is interesting to note that the biblical account of the Fall associates moral responsibility with the act of knowing. Doing and knowing are inextricably intertwined, much as Jonas has argued.) The point is made that anytime a value such as progress is absolutized, there is a tendency to justify unethical means in order to accomplish the end goal. In the process people may be simply used as sources of data unless there are limits set.

The two major underlying issues that concern opponents of human engineering research/application are fear of: (1) political control and abuse; (2) the alteration of human beings in ways which violate human integrity and dignity.

With regard to political abuse of human engineering knowledge, opponents usually point to the medical-political abuses of Nazi Germany. The scientific and medical community participated in the abuse of knowledge for what they considered a higher goal, i.e., human perfection. In response to claims that this could only happen in a dictatorship, opponents point out that the vast majority of contemporary governments are dictatorships. Furthermore, the pressures of overpopulation, famine, and economic difficulties may force many remaining non-dictatorships to become so in the future. In addition, historical experience with eugenics in the United States raises strong suspicion that a mixture of science and prejudice led to the passage of sterilization and miscegenation laws by 30 states.[16] Thousands of people were sterilized in the early 1900s due to such abnormal traits as drunkenness, criminality, and feeblemindedness. Sterilization laws still exist for the mentally retarded, and have been reconsidered for potential criminals

with an extra Y chromosome. The marriage restriction and immi-
gration laws were aimed at people of different racial groups
and immigrants. They were partly inspired by warnings about
contamination of American genetic stock. Opponents point out
that many scientists are once again using such concepts as
genetic load to justify their proposals. A bill recently drawn
up by the Chicago Bar Association would require mandatory genetic screening before marriage. The bill did not list or give
guidelines for defining defects. That was to be left to local
physicians or health officials, without legislative or public
action. The goal was to eventually reduce the number of non-
productive members in the society.[17] This, of course, could be
used to justify extensive political abuse.

Opponents of human engineering would generally concede that
most scientists are not attempting to be malicious or oligar-
chical elitists. However, they point that scientists are not
free of the desire for power or of political persuasions. They
are human beings as well as scientists and could, in some cases,
be persuaded to misuse their findings for ideological purposes.

Secondly, opponents argue against human engineering because
it poses a very real danger to the integrity and dignity of
human beings. They point out that the new genetic and psychological technology makes possible the apparently non-coercive
control of persons, for example. That is, a person produced
through various eugenic techniques would not be aware of being
genetically programmed, and therefore partially controlled by
the genetic programmers. Those subjected to various electrical,
chemical and conditioning interventions might not be aware that
they are under external control. Further, they might actually
prefer the controlled state if it is found as pleasurable as
some reports of psychoactive drug use and electrical stimulation
make the experiences sound.

Ironically, then, the fundamental issue on both sides of the
human engineering debate is *control*. Many scientists don't
want their activity controlled; the public citizen doesn't want
his life controlled! The irony is, of course, that

> .... our own in-groups, the groups with which we per-
> sonally identify ourselves never manipulate.... manipula-
> tion is always an act of somebody else.[18]

Both sides see the other as manipulative and controlling. The
challenge is to find an ethically and socially suitable framework which protects reasonable freedom for both scientist and
citizen.

15

Within this context several specific issues have emerged for intense discussion. These include: (1) informed consent; (2) protection of the institutionalized; (3) review of proposed research; and (4) definition of abnormality.

The *informed consent* issue basically centers upon the right of the individual to privacy and the exercise of political choice for purposes of self-protection versus the scientific right to know. It is often argued that the latter right frequently necessitates the use of procedures which at least partially obscure the research purposes in order to obtain accurate information. Opponents of this view argue that:

> Human experimentation for whatever purpose is always also a responsible, non-experimental definitive dealing with the subject himself. And not even the noblest purpose abrogates the obligations this involves.[19]

A host of controversy has been raised over the meaning of the words: is it ever possible for a layperson to be genuinely informed of technical procedures in such a way that he can reasonably assess personal risk? Is it possible to get true consent from the cognitively defective?

A recent study suggests that a majority of subjects involved in biomedical and behavioral research felt they were given understandable and accurate information, and that their participation was voluntary.[20] They also reported that consent forms were frequently written in difficult sentence structure that made them hard to read. A high percentage of those involved felt that they had directly benefitted by their participation.

Despite these recent survey results it seems that the construction of guidelines which will safeguard individual privacy and integrity, while permitting non-intrusive scientific investigation will be a continuing concern.

Related to the issue of informed consent is the *protection of the institutionalized* and minorities. The institutionalized groups under primary focus currently include prisoners, the mentally retarded, and those in mental hospitals. The issue of minority rights becomes involved because of the fact that the majority of prisoners are racial minority members, and those in all three types of institutions tend to be from lower socioeconomic groups. In addition, revelation of the procedures used in the Soviet Union to quiet dissidents by declaring them mentally ill, hospitalizing them for years, and subjecting them to abusive "rehabilitation" procedures, makes the issue a live one. Experimental medical procedures are typically conducted first on such populations. The distinction between therapy and experimentation frequently becomes murky. A host of related questions are also raised. With regard to prisoners: should they be subjected to rehabilitation procedures that will effectively help them be returned to society as non-violent, non-destructive, law-abiding procedures, even if this is against their will and they have been incarcerated for violent offenses? With regard to the mentally retarded: who is the most compassionate advocate for a mentally retarded child or adult? In general, how can the rights of religious, racial, sexual, and political minorities be protected if their beliefs and behaviors seem to be counter to the "public" good, and human engineering procedures might be employed which could return them as "compatible" members of society? Especially, how can this be safeguarded in non-democratically governed societies which could be expected to employ such procedures when they become available. Indeed, recent revelations suggest that the United States has been actively involved, through the Central Intelligence Agency, in training dictatorial governments in the use of sophisticated torture methods, which are then hideously used against political prisoners. Amnesty International has shown that in a horrendous number of cases individuals are incarcerated and tortured for years for no reason except whim. What is to prevent the abuse of human engineering procedures in this way? For the institutionalized, how far should individual rights be extended?

A third, and related, issue is the *review of research*. Who should review scientific research and decide whether or not: (1) it is ethical; (2) it should be performed. Until recently these decisions were considered either the sole province of the individual researcher, or subject only to a review committee of professional peers. Groups of scientists such as Science for the People, concerned about the social and political implications of science, have emerged on the national level to deal with these questions. Local, *ad hoc* groups such as the Massachusetts Advocacy Center Task Force on Children out of School have also begun to spring up and challenge the legitimacy of research in a given community that has not had community approval. The perspective of such groups is that:

> From the community point of view, it is also clear that peer-review of scientific research is inadequate for protection of the public interest. Peer review is like asking the CIA to review its own activities. There have been times when human research review committees have involved some community people, along with professionals in reviewing the human, legal, and ethical implications of research, but they have had a minority status. They

17

have not had the kind of clout to turn around and be
a force in those situations.[21]

Furthermore, these groups are not only asking to evaluate the
ethical nature of the research, but insist that:

> even after the scientific and ethical questions are
> answered, that doesn't mean that the research should
> be done. There's another set of social and political
> questions. It may be ethically acceptable, it may be
> scientifically valid, but it still may not be worth
> doing- because it's too expensive, because it's a bad
> way to use social resources, or because it offends
> someone in the community who has a goal that's more
> important than science.[22]

An increasing number of persons is, then, demanding that
experimentation involving human subjects or with significant
human implications not only involve more rigorous institutional
or peer review, but that it also be subject to public review.
Even laws have begun to emerge in certain communities (e.g.,
Cambridge, Massachusetts) which prohibit or restrict certain
human engineering-related research. On the other side are
scientists who argue that although the motivation behind such
proposals is commendable, the proposals are bound to have
disastrous effects for science and society. They argue that:
(1) non-scientists are not adequately enough informed to make
judicious decisions about scientific research; (2) that there
are an almost infinite number of groups who could claim to
represent a community; involving such *ad hoc* representation
could delay research so long that it would have to be aban-
doned or it would subject the scientist to the continuous
threat of suits; and (3) this process politicizes science un-
necessarily, and threatens its objective pursuits. The
strength of the arguments depend in part on the kind of public
review being recommended. Politicization will probably be less
if a (perhaps majority) number of public citizens are placed
on established review boards, than if research is reviewed in a
court-like procedure where testimony is received and decided
upon by all interested community groups.

A fourth issue that is being raised by some is the *definition
of abnormality*. The seriousness of the issue is raised in reference to the already mentioned Soviet practices. Should the
definition of disease include social deviance as the United
Nations suggests?

It is clear that:

18

once one chooses to consider the condition a disease
condition, the description of the "facts" of the
matter changes... The individual who is the focus of
concern becomes a "patient", his judgments concerning
the risks....become "rationalizations", his refusal
....becomes an "inability to function."[23]

Recent research by D. L. Rosenhan[24] revealed the startling
effects in the apparent perception and resultant interaction of
psychologically normal persons who, as part of research, were
admitted to a hospital as "patients." When is a genetic defect
abnormal, since all of us carry defective genomes? When is
violence normal and abnormal? If social deviance is considered
a disease, what is to prevent the politicization of biomedical
and behavioral intervention?

VALUE IMPLICATIONS

In addition to these specific issues, human engineering technology raises a number of important questions about traditional
Western values. It has fundamental implications for our conceptions of the individual, the family, and the state.

For example, cloning, genetic surgery, ectogenesis, and
psychosurgery raise difficult questions regarding human nature.
What does it mean to be human? Does psychosurgery create a new
person? Is a test tube baby a human being? Does a person on
drugs or using ESB for pleasure become less of a person? Do
these procedures violate human dignity and integrity? Do fetal
research and abortion constitute murder?

The value that is placed upon the individual is of extreme
significance in the formation of ethical guidelines. This is a
philosophical presupposition. If people are seen as individually valuable, ethical rules will tend to focus on the safeguarding of individual rights. Any other view will tend to subordinate individual good to the good of the state, or the good of the
greatest number. The kind of commitment that is made on this
issue will largely determine which side of the private right
versus the public good issue a decision-making body will come
down on.

Likewise, human engineering procedures raise basic questions
about the nature of the family. What will asexual reproduction
through cloning do to the husband-wife relationship? Moreover,
what will be the nature of the relationship between a child born
through *in vitro* fertilization and implantation in a mother surrogate be with the family? Who are his parents? What will his

psychological reactions be at the point of identity formation? Will amniocentesis and genetic screening strengthen or weaken family bonds?

What is the role of the state in response? Is it to guard the freedom of individuals? What if, in the face of overpopulation, famine and staggering medical costs, the expense of giving the freedom to procreate defective offspring is too much? What is too much? Should the state then restrict certain rights currently perceived as fundamental, in the interest of the common good? How can we be assured that the technology will not be politically abused? How will the government decide issues of legal identity in cases such as those mentioned above?

To what extent will we eventually consider these techniques as essential for the attainment of such widely accepted values as progress, security (personal, family and national), happiness, and life? Will they become rights for individuals? That is, will people be able to demand drugs, ESB, or genetic surgery in the name of freedom for the sake of pursuing happiness? Should they become individual rights? Should the government be allowed to invoke human engineering for the good of the people as a new kind of benevolent despotism? If not, what values would prevent this from happening?

How will these techniques change our conception of human destiny and of God? Will religious beliefs decline as man becomes able to create and control man? Will human engineering be justified on the basis of traditional Judeo-Christian values of peace, joy, love, patience, gentleness and self-control?

How will decisions about guiding values be made? Who will decide which values are preferable, and which ones are antiquated? What will happen to those who do not agree with the value hierarchy? Rokeach, in his monumental study of values, finds relatively little consensus with regard to the ranking of terminal and instrumental values (with the exception of happiness and freedom).[25] For example, he has identified 14 developmental patterns from adolescence to old age. Value preferences vary with race, culture, age, educational attainment, occupation and ideology.

Are happiness and freedom compatible terminal values? Does the pursuit of happiness sometimes mean the decline of freedom? Or does the protection of individual freedom mean that certain human engineering techniques which might bring a form of happiness will have to be restricted in their development and application?

20

The choice of values is also critical in the formation of an ethical decision-making framework. Unless a purely utilitarian approach is taken, values provide the foundation for ethical decisions. Even utilitarianism is based on the primary value of happiness for the greatest number. The derivation of values, then, is intricately involved with the formation of ethical norms. R. W. Sperry has rightly pointed out that:

> any concept or belief regarding the goal and value of life as a whole, once accepted, then logically supersedes and conditions the entire hierarchy of value priorities at all subsidiary levels. Values at the ideological plane become ordered, and ethical issues judged, in accordance with the conceived ultimate goal and purpose of life as a whole.[26]

One's view of human nature and purpose, then, orders all of our value priorities, and the way in which we make ethical decisions. If we believe that the primary human characteristic is rationality, we will most likely adopt some version of a deontological ethical model. A view of human nature as irrational is likely to lead to an ethical system that focuses on: (1) feelings--"if it feels good then it's okay," or "if I want it, it's alright to have it"; (2) intentions--"if I mean to do good what I do is right," or (3) outcomes--"whatever makes the most people happy is the right thing to do,"[27] A rational view of human nature will tend to result in a focus on normative ethics or ethical principles. The second view will tend to result in a focus on teleology or outcomes.

The demise of the dominant Judaeo-Christian religious system has resulted in a view of persons as predominantly biological with an emphasis on physically-oriented values, although the insurgence on Eastern religion in the United States and evidences of dissatisfaction with materialism as a guiding value may be signalling a return to spiritual values of a sort. The demise of Christianity in Western civilization has resulted, however, in ethical pluralism (or perhaps accompanied it). There is a lack of moral consensus. Modern man is unsure about the values which should guide ethical decisions. Up to this point, the contemporary god of science has seemed to foster skepticism and the absence of agreement, rather than the kind of ethical leadership some claim it is capable of deserving.[28] Unless an alternative is found, this decline in the concept of higher authority and moral consensus can only lead to the construction of an *ad hoc* political/judicial ethic that may or may not be moral.

21

Because of the importance of our view of human nature in the construction of ethical decision-making frameworks, I would like to digress briefly to investigate contemporary understanding of human nature.

HUMAN NATURE

At the outset it should be emphasized that there is no unanimous agreement about the nature of man.[29] There are several different philosophical, biological, psychological and theological models. Each emphasizes different aspects. The difficulty with most of these approaches is a tendency to oversimplify. The result is a truncated, incomplete, inadequate view of human nature. Above all else, human beings are complex. They cannot be accurately reduced to single concept, or single level explanations.

I would like to try and draw out some common agreements about human nature from these varying models. I will attempt to begin with those characteristics which are most widely agreed upon and move to those where there may be disagreement.

First, human beings are *multi-dimensional systems*. Man is a complex system of interacting capacities and needs. Human nature is not reducible to one or two characteristics. It is the *total system* which makes the human being unique, rather than isolated characteristics. Decisions made from a lesser conception of human nature will violate human dignity and integrity. It makes little sense to talk about human nature as purely psychological, or purely biological, for example. Human beings are an interlocking system of biological, psychological, social, and (for some) spiritual needs and functions. What occurs in one part of the system affects one or more of the other system components.

Human beings are *developmental and dynamic* in nature. They are not structurally or psychologically static. Rather, they are marked by generally orderly change which comes about through environmental encounters and biological events. Part of successful human functioning is the ability to adapt or change, to be appropriately flexible.

Human beings are *information-processing* organisms. The nervous system of the human being is constructed in such a way that incoming stimuli are organized into patterns and interpreted. A correlate of this information-processing capacity is the fact that human beings think, perceive, feel and behave symbolically. That is, humans are able to use 1-many language

22

rather than 1-1 signs. The capacity to abstract and form concepts liberates the human being from perceptual boundedness. The concepts which are formed influence all areas of human responding. The human cognitive capacity is far more complex and non-stereotypic than any other organism.

Human beings are *feedback-dependent*. They alter responses according to the kind of information and reinforcement received about present or past behavior.

Human beings are *historical*. Memory is one of the fascinating aspects of human nature. It is at least partially dependent upon the ability to think abstractly. One result of historicity is a sense of self-continuity or identity over time. Each of us is normally able to recognize ourselves as the same person regardless of our age. Historicity also has significant implications for the nature of interpersonal and intergroup interactions.

Human beings are *interpersonal*. A variety of psychological studies indicate that the need for positive relationships with significant other persons is a fundamental need. Our daily living centrally involves social psychological processes such as social comparison and modeling. A correlate of this characteristic is the need to love and to be loved. Positive self esteem is essential for healthy psychological and interpersonal functioning. This is rooted in the receipt of adequate social feedback which indicates that one is accepted, belongs and is valued.

Human beings are *purposive* or goal-oriented. Whether this is shaped by past reinforcements or abstract conceptualizations of what one thinks they might like to do on the basis of non-personal experience matters perhaps little. The fact is that human beings who seem to have no purpose or goals toward which to strive function less well and frequently develop serious difficulties.

Human beings are *self-conscious*. This is obviously dependent in part upon the ability to symbolize. The cause of self-consciousness is unknown. Every human being is able to self-reflect, however, if he possesses a minimal level of intelligence making it possible for abstraction to be carried on. This means, of course, that the self-reflections are able to influence human functioning as much as other stimuli.

Human beings are *rule-making* or *moral* in nature. Every human society has some kind of rules or expectations, the transgression of which is punished. Human beings naturally talk in terms

23

of oughts and shoulds with the meaning of obligation being conveyed. They are concerned with questions of right and wrong.
From early childhood they utilize a concept of equity of fair-
ness in social interaction that carries the meaning or ethicity.
In most societies (if not all) it is "right and good to be honest, generous, kind and brave, especially if in doing so one
brings hardship upon oneself."[30] A correlate of man's moral
nature is his responsibility for the effects of his responses,
and, according to most religious systems, his imperfection or
sinfulness.

Human beings are *volitional*. That is, they rebel against
the absence of choice and require the opportunity to exercise
choices in at least some areas of their lives. Psychological
research on reactance indicates that this sense of autonomy is
a vital component for Western man. There is no reason to believe, in the historical light of overthrow attempts of dictatorships, that this need is different in other countries.

Ironically, it is the capacity, indeed the necessity, for
choice areas which confronts man with the dilemmas of bioethics. Choice made on the foundation of finitude and self-
centeredness creates the problem.

Human beings are *emotional*. They are capable of expressing
a wide variety of feelings ranging from joy to hatred. Un-
checked emotionality typically leads to distortion and dysfunc-
tion, while the comparative lack of emotional expression is
usually seen as indication of pathology.

Human beings are *finite*. They are able to handle and process
meaningfully only limited amounts of information. They are
unable to project without error into the future because they do
not know enough; they are unable to accurately anticipate
changing events and parameters. They create negative situations as often from misunderstanding as from evil intention.
Human beings are not omniscient.

Human beings are *transcendent*. That is, they idealize and
ask ultimate questions about the meaning of life and death.
They seek to improve themselves, and to not be limited by the
contraints of the present. They engage in a religious quest
to move beyond themselves and to place their trust in some god,
person or cause that is bigger than themselves.

ETHICAL DECISION-MAKING MODELS

Ethical decision-making models can be classified in several
ways. They can be seen as rule-oriented, outcome-oriented, or

motive-oriented. *Rule-oriented*, or deontological models stress the need to determine basic principles by which a decision can be properly made, regardless of situation or results. Examples of this approach are the Natural Law approach, and Kantianism. *Outcome-oriented*, or teleological models emphasize the end or result of an action as the final ethical criterion. If a given action will produce the greatest amount of good for the most people it is judged to be ethical. The outcome-oriented ethic is pragmatic or utilitarian, while the rule-oriented model is an ethic of principles.[31] The *motive-oriented* model makes ethical evaluations on the basis of the intention behind the action. Regardless of the outcome, and regardless of whether other rules are broken, an act is judged ethical if it was done with a motive of love (desire to bring about beneficial results for the other). This approach is commonly known as situationism.

From a normative standpoint, ethical alternatives can be classified according to their view of universal norms. Geisler[32] suggests that there are six such alternatives: (1) antinomianism; (2) generalism; (2) situationism; (4) non-conflicting absolutism; (5) ideal absolutism; and (6) hierarchicalism.

Antinomianism is the view that there are no meaningful ethical norms at all. According to antinomianism there are no objective norms for making a judgment; acts can only be subjectively "justified". *Generalism* is the view that there are no universal norms, but there are moral principles which are generally valid. These general norms should be adhered to under normal circumstances, but the greater good or another overriding principle may occasionally necessitate violation of a specific rule. Such violation is not seen as unethical. *Situationism* holds that there is one universal norm, love. It is a form of absolutism but is so classified because a wide variety of otherwise "unethical" behaviors may be justified, depending on the ends that are operating in a given situation. If the transgression is done selflessly, out of love, the act becomes moral, according to this approach. *Non-conflicting absolutism* argues that there are many valid universal norms. Properly understood, however, these norms do not really conflict. There is always a third way, or creative ethical alternative. If immoral consequences follow from adherence to a universal principle, the adherent is absolved of moral responsibility. *Ideal absolutism* suggests that there are many conflicting universal norms. Violation of any such norm is wrong, but these violations can be ranked as more or less evil. Although evil is unavoidable, transgression is excusable, particularly if the lesser of the evils is chosen or there is confession. Finally, *hierarchicalism* says that, while there are many universal

ethical norms, they are not intrinsically equal. They can
be ordered so that when two norms come into conflict the
higher norm is to be obeyed. Hierarchicalism hold that norms
are universal within their context; there are exemptions, but
not exceptions.[33]

Another approach to the study of ethical decision-making
models is to classify them according to the source, or authority
from which they derive their ethical guidelines. Using this
approach we can classify them as: (1) inductive; (2) deductive;
(3) transductive.

The *inductively* derived models are situation-centered and
derived. They are built out of social consensus or from empirical induction. They are not derived from some notion of higher
principles or universals. Utilitarian, or outcome-oriented
models, fall into this category. Rule-utilitarianism, for
example, attempts to construct guidelines for generally maximizing happiness out of reflected-upon consequences of various
acts in similar situations. Related to this approach is
Kieffer's[34] suggestion that a societal ethic should be developed by means of a public democratic process; the ethic would
be inductively built from social consensus. Another inductive
approach is that which has been advocated by a number of
scientists.[35] These individuals argue that the fact-centered,
objective nature of science provides us with the best basis for
constructing an ethical system harmonious with reality, as
understood in nature's grand design.

The classical *deductive* model is, of course, Kantianism.
Kantianism holds that an act is moral only if governed by a
rational moral principle, or "good will." A moral act is based
on a sense of duty, and its morality is independent of the
results. The supreme principle of morality according to Kant
is that "I ought never to act except in such a way that I can
also will that my maxim should become a universal law." Kant
argued that some principles cannot be universalized without
involving logical contradictions, while others are psychologically contradictory (e.g., to wish eternal unhappiness upon
oneself is logically but not psychologically possible).
Kantianism holds that there are perfect duties which are never
right to violate, and imperfect duties which are not constantly
or universally binding. The Kantian views makes a fundamental
moral distinction as well between persons and things; persons
possess intrinsic worth and must not be treated merely as means
to an end.

A contemporary expression of neo-Kantianism is known as
structuralism. The argument is basically that DNA provides a

26

source of survival and adaptational constraints which limit the range of acceptable ethical systems, and provides a basis for avoiding egoistic utilitarianism. The structuralist approach holds that moral judgments arise by means of a generative process which reflects an innate, deep ethical structure that is universal and genetically based.

These universal deep structures would be more or less equivalent to Kant's concept of the categorical impera- tive... However, the neo-Kantian feature of the structuralist approach to morality is that it would posit a more complicated, or transformational, relation between the universal categorical imperative and par- ticular moral judgments that the direct connection evidently envisaged by Kant.[36]

Structuralism draws support from the approach to linguistics championed by Noam Chomsky who argues that linguistic patterns are innate and based on the same universal grammar. In addi- tion, insights from biology with regard to transformation of light rays into a visual percept, and some of the assertions of the new sociobiology about the possible genetic foundations of certain moral traits such as altruism [37] are supportive.

A third deductive approach is that of natural law. Natural law theories insist that the good is something objective in contrast to utilitarian theories, but also ethics must be grounded in concern for the human good.[38] They are based on rationalism. The natural law ethic has arisen primarily out of the Catholic religious tradition, and specifically from the philosophizing of Aquinas. It proceeds from moral principles to practical applications. The basic precept of natural law is to do good and avoid evil. Several subsidiary principles, such as the principle of double effect, are used to interpret the general precept. The law of double effect suggests that actions can be justified only if evil is not the means of producing a good effect, the evil is not directly intended, and there is a proportionate reason for engaging in an act.

Finally, within the Judeo-Christian religious tradition the revelational model has been propounded. This is related to the natural law model, but tends to be more representative of Protestantism. In addition to the general revelation of nature, specific moral guidelines and commandments have been given by God to His people. Adherence to these guidelines is presented as the source of blessing or disaster for that people. Although the guidelines are specifically given to this group it appears that God's punishment has been meted out to others who do not follow them. This view is held as at least necessary (though

27

perhaps not sufficient) to proper ethical decision-making by contemporary evangelicals.

In summary, the deductive model holds that ethical principles are derived either from human evolutionary and genetic structures or from a Higher Justice (usually seen as residing in God). According to this view there are universal norms, not simply cultural norms or situational behaviors.

Two examples of the *transductive* model are situationism and hierarchicalism. The transductive model attempts to establish universal ethical principles, while at the same time interpreting the rules at least partially in the light of a given situation.

PROBLEMS

Each of the models and their particular expressions has some difficulties. Technically speaking, the inductive model is not predictive (that is, the morality is situation-dependent) and not moral (if moral philosophers are correct, the "is" cannot properly become the "ought" in a moral sense). One of the major difficulties with the deductive model is the general nature of the universal principles. Interpretation within a specific ethical decision-making situation is extremely difficult.

> So while consequentialist theories are, in general, *epistemologically* superior to deontological theories in that they enable us to arrive at unique solutions to moral dilemmas about how we should act, many would argue that such theories are, nonetheless, *ethically* inferior to deontological ones since they fail to provide a justification for why we should do our duty when duty conflicts with maximizing desirable consequences of an action.

The transductive model has difficulties as well. Central is the question of weighting--to what extent should principle or situation determine the specific decision. How does one arrive at a hierarchy that is non-arbitrary in the case of hierachicalism? How does one maintain a concept of love that makes the notion of morality meaningful in situationism?

Amazingly, proponents of sociobiology, structuralism, and the conscious determinants model seem to agree that the shape of the general values which will guide ethical decision-making if their model is adopted will be very similar to traditional models. The feeling is "the golden rule" in its general form is common to most cultures as a moral backbone, and that, for

28

whatever theoretical reasons this is so, this principle and its correlates will and should be maintained.

Beyond that very general point of agreement, however, contemporary ethics seems to be at an impasse. Without value consensus the likelihood of developing a decision-making model that is commonly accepted is small. As we have already mentioned, the probable result will be a patchwork mixture of guidelines that are political and judicial in nature. The practical guideline employed could conceivably become a non-ethical value such as economic benefit/cost.

THE MEETING PLACE

The relationship between the evangelical Christian and secular scientific communities has been marked by discord, mutual suspicion, and mutual rejection for much of the past 250 years. Although many outstanding scientists have been personally committed to the Christian faith, many scientists have regarded such belief as either completely irrelevant to science, or as a dangerous hindrance to the objective pursuit of knowledge. The conflict is on presuppositional grounds. Christianity is revelational, deductive and historical in approach. Science is empirical, inductive and experimental. Christianity has focused on the future; science has dealt with the causal past. On the other hand, many evangelicals have viewed science and scientists with a mixture of suspicion, fear and open hostility. Science has challenged traditional ways of thinking and doing. It has threatened fundamental conceptions of human nature and purpose. Its empirical and inductive nature has led to widespread questioning about the existence of God.

Until very recently the perceived need for scientists to consider "non-objective" value-related issues as integral to both the investigation and implementation of science was not accepted. Because of the increasingly understood link between research and application, pure science and public policy, questions of ethics and values have become important. This has been especially true in areas of science involving human experimentation and implementation.

One's treatment of human beings intrinsically involves ethical and value implications. As a result, the prestigious National Science Foundation has established an Ethics and Values in Science and Technology (EVIST) program.[40] The EVIST program is to promote scientific, professional and public understanding of ethical problems associated with science, the impact of changing ethical and social standards on the activity of science,

29

and the processes which lead to the generation and resolution of conflict between scientific and social groups.

It is exactly at this point that the evangelical scientist and community should be actively involved in helping to develop an ethical/value framework which recognizes the legitimate domain of scientific investigation but also grapples with constructing appropriate guidelines which protect individual freedom and dignity. The evangelical has both a basic commitment to the pursuit of truth, and a personal commitment to a historically significant system of values and ethics which needs to be unpacked and applied to human engineering issues. Governmental and scientific bodies which ignore this rich ethical history can only be short-sighted and prejudiced. On the other hand, evangelicals who think that only their way will lead to the safeguarding of the human race are arrogant. The truth is, as has already been mentioned, that the traditional values derived from our Christian heritage are unlikely to be significantly changed. There is a common historical pool of consensus about certain of those broad values (e.g., individual freedom) which provides a place of much needed dialogue between those who have characteristically removed themselves from moral questions, and those who have often removed themselves from the legitimate claims of science. Concern about the ethics of human engineering experimentation and application provides a common meeting place for science and Christianity. It should be noted, as well, that other value-oriented groups (religions) should have equal opportunity to have input.

SOME SPECIFIC CHRISTIAN CONCERNS

Commitment to a conservative Christian theological position has characteristically carried with it some specific concerns about human-oriented research. Briefly, several of these are:

1. Protection of individual integrity and dignity. The concern here is that basic human nature not be violated by research procedures or applications. Christians are especially, though not exclusively, concerned about the implications of these procedures for an individual's relationship with God. Does it enhance or impair the person's ability to relate to, worship and glorify God? Does it enslave or enhance the individual's capacity to make choices?

2. Compatibility with creation ordinances. What implications do the procedures have for marriage, family identity, morality, fidelity, loyalty? Does genetic engineering violate the Creation boundaries; that is, is man usurping God's role?

30

3. Rehabilitation versus retribution. What kind of response should be directed toward those who have violated basic legal and moral commandments? Should they be rehabilitated by means of some of the human engineering procedures, or simply receive retribution as in the Old Testament? How far do the rights of offenders extend?

4. Specific ethical guidelines. What does the Bible have to say by way of principles that will give guidance in the search for an ethic of human experimentation? The Bible does not seem to be against change. It does not even absolutize individual rights; if prior choices negated God's commandments, for example, both choice and life could be legitimately taken away. Specific extensions of general principles are a major concern.

5. Manipulation and dialogue. To what extent are human engineering procedures manipulative and dehumanizing? The Scriptural pattern seems to establish dialogue or mutual interaction as the means for change. Should manipulative procedures that are used in some research be partially or totally opposed? What are the options for investigation with a dialogic model? To what extent does a given human engineering procedure make it possible for long-range control of an individual by an external agent? In this regard, the Christian community has tended to favor procedures which are reversible, or can be undone, to those which are unchangeable and permanent in effect.

6. Improvability. Does God set any limits on biological or psychological movement toward perfection by non-spiritual means? Should such techniques be used to aid in the development of positive characteristics commonly associated as fruits or products of the activity of God's Spirit?

7. Compassion and ambition. Does our responsibility to human beings imply the taking of risks which may bring good or may bring bad consequences? Because of possible risks can we oppose the development and application of procedures which may relieve suffering and aid human functioning? What are the demands and limits of compassion? How can we protect ourselves from the confounding of compassion and ambition?

31

STRUCTURE OF THE BOOK

MODIFYING MAN follows the basic structure used for the International Conference on Human Engineering and the Future of Man. The book is divided into eight sections. In the first seven sections a specific set of issues or human engineering procedures are investigated. First, a major paper is presented by a noted authority. The major addresses give a brief overview of specific human engineering procedures, and then focus upon the ethical and value implications of these procedures for individuals and society. Then, responses to each major address are given by an interdisciplinary panel of evangelical scholars. Among the disciplines represented by respondents are genetics, theology, psychology, philosophy, sociology, law, psychiatry, and biochemistry.

The second section deals with some of the general philosophical and practical issues raised by human engineering technology. Section III focuses on theological issues of concern to those wrestling with the genetic and psychological modification of human beings. The fourth section presents an overview of the genetic intervention techniques and the value implications associated with them. Section V focuses on behavior control by means of psychosurgical procedures. Section VI is concerned with the use of psychological control, and its potential mass applications. In the seventh section a look is taken at the formation of public policy, and, in particular, the place of dialogue between the scientific, legislative and religious communities.

The final section of the book presents an interpretative summary of the Conference based upon formal presentations and dominant positions taken by conference participants in small groups This is followed by the Conference Commission statement, "Evangelical Perspectives on Human Engineering," developed subsequent to the main portion of the conference, but intended to reflect Commission members' assessment of attitudes expressed by members of the small groups which they led during the conference. The statement is preliminary and exhortative in nature, rather than final and definitive. It is intended to stimulate dialogue between the scientific and religious communities, and to promote serious consideration of these issues within the religious community. Finally, a brief description is given of the organizations that co-sponsored the conference.

32

FOOTNOTES

[1] George H. Kieffer, *Ethical issues and the life sciences*. Washington, D.C.: American Association for the Advancement of Science, Study Guides on Contemporary Problems (Test edition), 1975.

[2] *Assessing biomedical technologies: An inquiry into the nature of the process*. Washington, D.C.: National Academy of Sciences, Committee on the Life Sciences and Social Policy, Assembly of Behavioral and Social Sciences, 1975.

[3] Kieffer, *op. cit.*, p. 45.

[4] *Ibid*, p. 52.

[5] Tabitha M. Powledge, Recombinant DNA: The argument shifts. *The Hastings Center Report*, April 1977, p. 18-19.

[6] Kieffer, *op. cit.*, p. 44.

[7] Elliot Valenstein has provided an excellent overview of psychosurgical effects and the difficulties associated with accurate psychosurgical evaluation in *Brain Control*, (N.Y.: John Wiley & Sons, 1973).

[8] Joel Fort and Christopher T. Cory, *American drugstore*. (Boston: Educational Associates, 1975), p. 6.

[9] *Ibid*, p. 20.

[10] I am indebted to George Kieffer, *op.cit.* (151-154), for his discussion and classification of psychoactive drugs.

[11] Fort and Cory, *op. cit.*, p. 6.

[12] Robert G. Heath, Electrical self-stimulation of the brain in man. *American Journal of Psychiatry*, 1963 (120), p. 571-577.

[13] Kieffer, *op. cit.*, 136-142.

[14] Hans Jonas, Freedom of scientific inquiry and the public interest. *The Hastings Center Report*, Vol. 6 (No. 4), August 1976, p. 15.

[15] *Ibid*, p. 16.

[16] Frederick Ausubel, Jon Beckwith, and Kaaren Janssen, The politics of genetic engineering: Who decides who's defective? *Psychology Today*, June 1974, p. 40.

[17] *Ibid*, p. 38.

[18] David O. Moberg, The manipulation of human behavior. *Journal of the American Scientific Affiliation*, March 1970, p. 14-15.

[19] Hans Jonas, Philosophical reflections on experimenting with human subjects. *Daedalus*, Spring 1969: *Ethical aspects of experimentation with human subjects*.

[20] Research subjects report they are adequately informed; cite immediate and future benefits of participation. *Institute for Social Research Newsletter* (University of Michigan). Summary of report prepared by Robert Cooke and Arnold Tannenbaum for the National Commission for the Protection of Human Subjects of Biomedical and Behavioral Research.

[21] Hubert Jones, conference participant. MBD, drug research and the schools: A conference on medical responsibility and community control. *The Hastings Center Report*, June 1976, Special supplement, p. 7.

[22] Robert Michels, conference participant. MBD, drug research and the schools: A conference on medical responsibility and community control. *The Hastings Center Report*, June 1976, Special supplement, p. 15.

[23] H. Tristam Englehardt, Fear of flying: the psychiatrist's role in war. *The Hastings Center Report*, February 1976, p. 21.

[24] D. L. Rosenhan. On being sane in insane places, *Science*, January 19, 1973, p. 250-258.

[25] Milton Rokeach, *The nature of human values*. (N.Y.: The Free Press, 1973).

[26] R. W. Sperry. Bridging science and values: A unifying view of mind and brain. *American Psychologist*, April 1977 (Vol. 32, No. 4), p. 243.

[27] Some are suggesting that there is another naturalistic approach that is not utilitarian. We will consider structuralist ethics shortly.

[28] Sperry, *op. cit.*, p. 237-245.

[29] See "Indicators of Humanhood: A tentative profile of man," *The Hastings Center Report*, November 1972, 1-4 and "Four indicators of Humanhood - the enquiry matters," *The Hastings Center Report*, December 1974, p. 4-7, for discussions of this topic by Joseph Fletcher.

[30] Kieffer, *op.cit.*, p. 14.

[31] For a thorough discussion of Kantian and utilitarian ethics see the introductory chapter, "Ethical theory in the medical context," in *Ethical Issues in Modern Medicine*, edited by Robert Hunt and John Anas (Palo Alto: Mayfield Publishing Company, 1977).

[32] I am indebted to Norman Geisler for this classification of normative ethics in his book, *Ethics: Alternatives and Issues*, (Grand Rapids: Zondervan Publishing Co., 1971).

[33] Ibid, p. 19.

[34] Kieffer, *op. cit.*, p. 24.

[35] Sperry, *op. cit.*, p. 238.

[36] Gunther S. Stent, The poverty of scientism and the promise of structural ethics. *The Hastings Center Report*, December 1976, p. 39.

[37] Donald T. Campbell, On the conflicts between biological and social evolution and between psychology and moral tradition. *American Psychologist*, December 1975, p.1103-1126.

[38] Hunt and Arras, *op.cit.*, p. 40.

[39] Ruth Macklin, Moral concerns and appeals to rights and duties. *The Hastings Center Report*, October 1976, p. 36.

[40] *Ethics and values in science and technology: program announcement.* National Science Foundation, Office of Science and Society,(Washington, D.C. 1976).

PART II
PHILOSOPHICAL PERSPECTIVES

In this section Daniel Callahan argues that the central
ethical problem facing contemporary and future humanity in
the face of biomedical technologies is to find ways to live
with our finitude. As a result of finite power, Callahan
asserts, we are able to undertake acts of biological and
psychological control previously unthought of. At the same
time that we undertake these actions with good intentions,
we do not have adequate wisdom and foresight to anticipate
all of the potential consequences. We have the means to
make fundamental changes in our lives but we do not have the
capacity to know everything.

Callahan goes on to point out three other classes of
issues that are critical: (1) the need to rethink traditional values and definitions in the light of current
knowledge. Callahan believes that the traditional value
framework of the Judeo-Christian system is adequate in
general, but may have to be reinterpreted; (2) the process
of decision-making. Should new knowledge be made mandatory?
Do some individual rights have to be restricted for the
common good? Should basic research be subject to public
control? (3) establishing standards. Dr. Callahan considers
the emotionally charged issue of whether professional and
scientific standards should be established in part by the
public.

In response, psychiatrist David Allen suggests five
dimensions that must be considered in establishing the
ethical constraints of the decision making process. First,
he argues that there must be a thorough committment to
securing accurate information. Such information, according
to Allen, must be gathered from scientific investigation,
but also from other fields and from persons most likely to
be affected by a given ethical decision. Second, Dr. Allen
says that one's basic belief concerning the nature and
meaning of man will affect our decisions. He argues for a

conception of persons made in the image of God. Third, he
suggests that the kind of moral reasoning that is engaged in
must be carefully understood. In this section, Allen
surveys reasoning on the basis of utilitarianism and equality,
and evaluates these two approaches. Fourth, Dr. Allen
suggests that our presuppositions or loyalties will significantly affect our ethical decisions. Finally, he makes an
impassioned plea for compassionate and empathetic decision-
making.

Richard Spencer points out the biological bias that is
often implicit in considerations of quality of life. He
also questions the assertions that behavior modification
bypasses free choice and that control over aging would bring
total power. Dr. Spencer then argues that a proper concept
of history is essential for ethical decision-making. He
shows that the critical variable in human development
throughout history has not been biological quality but the
quality of relationships. Acceptability of human engineering procedures must be evaluated critically with regard to
implications for such relationships, according to Spencer.
The lessons of history also encourage caution and humility,
because they point out the degree to which our perceptions
and understanding are affected by culture and point in time.
Such tunnel vision can only lead to eventual disaster.
Progress must be evaluated carefully because our very
concept of progress is temporally and culturally shaped
and interpreted.

CONTROL TECHNOLOGY, VALUES AND THE FUTURE

Daniel Callahan
Institute of Society, Ethics and the Life Sciences

Human beings have always been a problem to themselves. On the one hand, mankind has been blessed with the gift of reason. On the other, that reason has led man to speculate about himself, both about his glories and possibilities, and also about his sins and follies. Whether it even makes any sense to talk about re-making man, or whether it is possible to so manipulate and change human nature that a different kind of being would emerge altogether is entirely unclear. But it is surely the case that human life is different now than it was in earlier centuries, and it seems to be changing all the time. It is not only the gradual accumulation of human technology which has caused such change in human life and culture; it is also, and most preeminently, the fact of human technology which has changed and continues to change without an apparent end in sight.

The technological powers of biomedical science are among the most significant of the twentieth century. One can hardly ignore earlier technological feats which led to nuclear weapons. But recently developing biomedical technology raises the newest and perhaps most difficult problems of all. Its great attraction is that it gives promise to coping with some of mankind's oldest enemies: disease and death, physical and mental impairment, and the fact that human life until very recently has been

---

DANIEL CALLAHAN was the founder of the Institute of Society, Ethics and the Life Sciences. Prior to that he was Executive Editor of COMMONWEAL and Visiting Professor at several Eastern Universities. He is a consultant on Medical Ethics to the American Medical Association, Judicial Council. He has authored or edited 14 books and over 100 articles. His recent books include THE TYRANNY OF SURVIVAL, ABORTION: LAW, CHOICE AND MORALITY, ETHICS AND POPULATION LIMITATION, THE AMERICAN POPULATION DEBATE, and ETHICAL ISSUES IN GENETIC COUNSELING AND THE USE OF GENETIC KNOWLEDGE (co-editor). In 1974 he was chosen as one of the 200 outstanding young leaders in the United States by TIME magazine.

thoroughly controlled, not only by the environment, but also by man's natural genetic and physical makeup. Whereas it was only a dream for earlier generations to think of doing something about the ills which afflict the body, the fact that we are often less than we wish to be, that we are often more controlled by our environment than we would wish now, our generation seems on the verge of being able to shape much of its own biological and behavioral future.

THE DILEMMA OF LIMITED POWER

The most prominent feature of the new biomedical technologies is that of power. Science is quickly granting us the power to change, manipulate, and transcend what seemed, to earlier peoples, fixed and firm limits. At the same time, the most important point is that the new powers are now, and will continue to be, limited powers. Nothing poses more difficult ethical and value questions than the possession of power which is not infinite, but has unknown boundaries. For then human beings must learn how to live both with the power they possess, and the limits of that power. We know that, whether it is a matter of technology, politics, or human relationships, power, if not used wisely, can be a source of enormous destruction. We are faced with a perverse paradox--the more strength we have, the more problems we create; the more power we have, the more in danger we are of being possessed by that power; the good we can do, the more harm we can also do. This is what makes life both enticing, and at the same time, enormously frustrating. We are finite beings, and our power is finite as well.

Let me illustrate some of the enormous dilemmas posed to us by the fact of incomplete power. The problem which most beautifully and poignantly presents the issue is the world population problem. Assuming that there is a problem, which I do, one must recognize that it is a result of the rapidity with which death-rates have been reduced around the world, that is not matched by equal speed in the reduction of birth-rates. Thus world population growth is doubling at a present rate of nearly every 37 years. Yet who could ever have forseen such a problem? Mankind is achieving one of its oldest dreams--not that of conquering death, but of staving it off to some extent--only to find that it has perhaps set the stage for an even worse incursion of death at a later date due to over-population. No one, I believe, would want to go back to a situation where death-rates rose, where children died in great numbers, where it was an amazing feat if one could survive to adulthood. But at the same time, how are we to achieve the kind of parity between birth and death-rates which will guarantee that the population of the world will not exceed its carrying power?

40

If the population problem poses one painfully obvious feature about having some power, but not total power, it also provides a fine introduction to what is surely one of the most important and amazing features of biomedical technology: the increasing power to control procreation. The development of modern contraceptives means that for the first time human beings have an almost decisive control over the number of children they have, and over the spacing of those children. That new power has meant that enormous changes are taking place in the parent-child relationship, the relationship between husbands and wives, and the relationship between the family and society. We do not yet know whether these changes in the long run will prove altogether beneficial, or what will be the mixture of benefit and harm.

If it is the case that people can now control the quantity of children they have, it is also increasingly the case that they can control the quality of their children. Through the increased possibilities of genetic knowledge, particularly the advent of screening for genetic disease and the use of prenatal diagnosis to detect defective children *in utero* people will be able, in many instances, either to avoid having defective children, or to have foreknowledge of the possibility that they will bear defective children. At this point, the immediate prospect is for a reduction in the number of genetically defective children. But even that possibility forces new dilemmas upon people: who among us can really know what kind of children we ought to have? Who among us can really speak with any final wisdom on the question of the genetic quality of children? That it may be possible in the distant future for people not only to avoid having defective children, but also to have greater control over the positive traits and qualities they would like to see in their children will raise even more difficult problems. The possibility for couples to select the sex of their children, certain to become much more widespread in the near future, will give humans a kind of control over the quality of their children simply not attainable in the past. But again, who among us can really predict whether we would be happier if we had more girls than more boys, or more boys than more girls, or some particular balance of both?

If we move away from the astonishing and troublesome problems of the power to control the quantity and quality of children to other new medical powers, we see issues no less appalling and difficult. It is now possible through medical technology to shape, manipulate and modify human behavior through a great range of techniques. These include psychosurgery, psychotropic drugs (which can either exhilarate, depress, or tranquilize people), and the psychological means of controlling behavior in a great.

41

variety of settings including schools, prisons, and asylums.

The essence of these new powers is that of bypassing free choice. The technology is opening up ways to get people to behave which were not possible when the only available means were those of irrational persuasion or direct physical coercion and punishment.

The fact that we can, with increasing precision, produce almost any desired emotional state, and manipulate memory and to some extent rational power, means that enormous power is now being placed in the hands of the medical and scientific professions as well as in the hands of society.

Yet it is by no means the case that this power is complete. People can and do refuse to make use of these new technologies. Courts often intervene to stop their use. There are serious reservations within the medical profession about the circumstances under which the power to modify human behavior and emotion should be employed at all. Moreover, it is now apparent that it is very difficult to modify human behavior and emotion without paying a certain price. Psychosurgery, perhaps the most dramatic of all the means of behavior modification, is by no means a well-developed science. We do not really know what the long term effects of psychosurgery are, nor do we have a clear idea of the conditions under which it is appropriately used. While there is considerably more knowledge of psychotropic drugs, they have not been developed to a point where they can achieve exactly the results intended without the possibility of some hazardous side-effects.

Once again, we have power, but limited power only, and of course limited knowledge to go with that limited power.

If we move to the end of human life, to the control of human death, we see a situation no less full of paradoxes and limitations. It is now possible to extend the life-spans of individuals, in some instances by the use of organ transplants. It is possible to keep people "alive" artificially on heart-lung machines, at least to keep respiration going and blood circulating through the body. It is possible through radical therapies, whether surgical or involving the use of drugs, to keep cancer and heart victims going in a way that was simply not even imaginable as long as fifty years ago.

And yet, because we have not been able to conquer death entirely but our power is still limited, we seem to have created for ourselves unique and very complex ethical dilemmas: Under what conditions ought people's lives to be extended artificially?

42

To what extent do we want to continue treatment, when such continuation may result in a life which is simply a misery to the person whose life we have "saved"?

If people could be kept alive indefinitely without aging we would have total power. But we do not have that kind of power, nor are we ever likely to have it. We will only have limited power. That is what we will have to live with.

All of the issues I have so far touched upon are, taken together, simply appalling. Our society does not agree on how they should be responded to. Nor is there total agreement within the scientific and medical community. It is, moreover, very difficult to see how we ought to go about thinking through these issues. Trends and rough systems of values will undoubtedly develop. These may be decisive in the way we handle the problems on a day-to-day basis. The central issue is to see if we can find our way through the problem of the use of limited power and come to solutions which do justice to the dignity of man while allowing society to function in a harmonious and wise fashion.

VALUES AND DEFINITIONS

There are a number of significant issues raised by biomedical technology. I have hardly exhausted the list. The first set of issues can be summed up with the following question: are traditional values and ethical principles still adequate? I mean essentially those which have come out of the general tradition of western ethics, including the Judeo-Christian system. This tradition has stressed individual rights and human dignity, the sanctity of life, and the need to care for people in times of illness and distress. Of course there has been enormous difficulty throughout the ages both in understanding these values and in trying to live up to their implications. Regardless, they have provided us with broad guidelines which have pushed our culture in one direction rather than another. These have led to a natural abhorrence of totalitarianism, cruelty toward human beings, the destruction and disposal of the weak, the aged, and the ill. Western values have led us, at least in theory, to take all human life seriously.

But do the new powers of biomedicine mean that we must rethink those values? My own view is that the traditional values still serve us adequately. The more difficult question is how we are to interpret some of those traditional values in the light of the new issues. How, for instance, are we to seek through medicine to protect human life and to minister to people in their illness, while at the same time not extending their lifespan beyond all

43

sense and reason just because we can do so? How are we to decide
about terminating treatment of a dying patient when it is very
difficult to even know what we mean by the "quality of life",
especially when it is being sustained only artificially? If we
can, to a great extent, control and modify human behavior, under
what circumstances should we do so? How are we to protect rights
in the process, and whose rights--those of the community at large
or those of individuals?

Our society has always been ambivalent about the question of
birth and procreation. On the one hand we recognize that it is
by no means a blessing that defective children are born. On the
other, we hesitate to define certain classes of human beings out
of the human community, simply because they were born with mental
or physical defects.

How are we to keep alive something which it took centuries for
our culture to develop--care and respect for the mentally defec-
tive--while at the same time, making use of our new scientific
powers to avoid the birth of many future defectives? Will there
not be a tension between our desire to love and cherish those who
are defective and our desire that fewer defectives be born? How
can we have it both ways?

A second large area of issues may seem trivial, but I believe
are critical. I am thinking here of the way we define some basic
terms. What, for instance, do we currently mean by "death"? Do
we mean the death of the brain, or the death of the heart, or the
death of the whole body? At one time, there was no problem about
this: people just died, and it didn't matter much whether it was
their brain or their heart one was talking about. Now, however,
because people can be sustained artificially and perhaps indefinitely, will we not be forced to be more precise about our meaning of "death"?

What do we mean by "human life"? This is an issue which looms
very large in the debate over abortion, and in the discussion of
the extent to which the aged and senile should continue indefinitely receiving medical treatment. The phrase "quality of life"
is thrown around with great abandon these days. Is there anything
to the notion, or is it simply a seductive phrase which will lead
to the elimination of the unfit and the undesirable? What does
it mean to respect the dignity of individuals? Does that respect
mean that people should have total control over their own life and
death--the right to take any drug they want, or the right to tell
a physician to quietly put them out of their misery, or the right
of a woman to have an abortion regardless of circumstances?

Even more broadly, we are going to be forced by these new issues

44

to cope afresh with the meaning of such old terms as health
and illness. The World Health Organization, in the late for-
ties, defined health to include social well-being. It also
included mental health as part of health in general. But what
could we possibly mean when we say that health should involve
social well-being? If we take that kind of definition seriously, we are really saying that all human problems and miseries--
political, cultural, and spiritual--come down to problems of
health. Does that make any real sense, or have we not stretched
the meaning of health to the point of sheer meaninglessness?

A very practical aspect of the way we go about defining health
and illness bears on the problem of the use of our medical powers
in this or that specific case. Are we to say that violently aggressive people who pose a danger to society are always "sick"?
Are they to be subjected to medical power to manipulate or modify
their behavior? If some people desire to choose the sex of their
children, are we to define that as a health problem, or as some
other kind of problem? More broadly, are all problems which the
medical profession deals with to be defined as problems of
health. As a practical matter, any national health insurance
plan will have to decide just what states of human life, and what
forms of behavior, are to be considered matters of health and
thus to be treated with government funds. That is a key political
issue, but one which takes us down to the very depths of our
thinking about what some fundamental concepts mean.

DECISION-MAKING

Another range of large issues raised by the new powers focuses
on decision-making. In earlier times, when medicine could do very
little for human beings, there were no decisions to be made. Nature simply took its course. Now, however, people have to make
choices about whether or not to use the new powers. If they are
to be used, the question is to what extent? How are we to go
about allocating power and responsibility for these decisions?
It is very easy to say that all choices should be left to the
individual, but clearly that is not, in many cases, a feasible
social solution. We are not likely to leave the question of the
use of brain surgery for the modification of human aggressiveness
in a prison setting entirely up to the private values and decisions of those who can perform this surgery. And yet on the other
hand, when it comes to the problem of abortion, the Supreme Court
has left that decision solely in the hands of individual women.
Do we have any kind of general theory, or theory in the making,
to deal with decision-making and locus of responsibility? So far
as I can see, we do not. Rather, we move from issue to issue with
little regard for general theory. We have no sense of how one
ought even to construct a theoretical solution to the question of
responsibility.

45

Before I speak directly about decision-making, let me make one point very clear. It is terribly easy in our society, which is divided by great value, ethical and religious differences, to think that all ethical questions come down to questions of "who should make the decision?" In many respects, of course, that is very true. But I think there are even more fundamental questions that need to be recognized. One can hardly even begin dealing with the question of decision-making until one has some central grasp of the ethical issues at stake in the decision to be made. We cannot, in other words, separate out the question of the process of making decisions, from the ethical essence of the decision to be made. A failure to recognize that point has meant that many so-called value or ethical debates have managed to avoid that very central ethical and value question at stake.

Let me now raise four major problems about the question of decision-making in the light of our new knowledge and power. The first problem is whether and to what extent the new knowledge and power we have should be made mandatory? If not mandatory in all cases, in which cases? For instance, it is now possible to screen people, either as children or perhaps before marriage, or perhaps when they are pregnant, to determine the possibility of their procreating a defective child. Should people be forced to have such knowledge? Or should they be free to remain ignorant?

Knowledge, it has been said, is power. But do we want knowledge imposed on people in all cases? Should there be circumstances under which they might be quite free to say "I do not want to know and I'm willing to take my chances"? Our society certainly does not allow people to remain totally ignorant in general, as our mandatory laws on the education of children demonstrate. Moreover, society might well say that it is simply irresponsible for people to deliberately remain ignorant about the fact that they bear a defective child. After all, society may, as much as the individual parents, have to support any defective children who are born. But what rights do we want to give society to override individual values and desires? Is the right to remain ignorant an absolute right, or one which the state might reasonably abridge if the state would be affected by decisions made in ignorance? If we force people into knowledge, can we force them also into action such as aborting a known defective fetus, or of refusing to allow people known to be carriers of dangerous genetic diseases to marry at all, or, if they marry, to procreate?

46

Another set of problems concerning decision-making can be put in the following way. Do we want to take away some traditionally cherished individual liberties in order to achieve socially desirable goals? For example, the right to procreate without interference from the state has become a well-cherished right. In the light of a desperate population problem, should that right be limited? We have cherished the notion that the individual's body may not be invaded against his or her will. But what if we could gain the release of someone from prison, by imposing upon that person unwanted psychotropic drugs or brain surgery? If the person would gain, and perhaps society as well, would that not be sufficient reason to go ahead with such procedures even if the individual in question said he did not want that to happen?

A third important set of problems concerning decision-making may be put in the following way. Since basic medical research can lead to such enormous powers for good or evil, should that basic research be subject to public control? Our tradition has said that the freedom of scientific reasearch ought to be absolute. Put another way, there are no restraints which should be exercised against the individual quest of the scientist for scientific truth. But, what if we see that certain types of basic research may well lead to great harm to the public? Should the public be consulted in those cases? Should the individual scientist be restricted if the public felt that the research was going to be harmful to its interest? A very important principle in our society is that those who are likely to be harmed or affected by anything should have some say in the matter; otherwise, their dignity and freedom may be violated without their consent. Do we want to apply that principle in the case of basic research which may produce both good and bad effects?

A particularly intriguing area for contemplation in this respect is human aging. There are many who are doing basic research in the biological processes of human aging. Are we prepared to give total blessings to this research, particularly when we have no idea whether it would be good or bad for society to see the average life span extended ten, twenty, or fifty years? Ironically, we will never find out unless the research goes forward. But we may also find that we were sorry we started in the first place.

ESTABLISHING STANDARDS

The fourth large problem area I would point to can be summed up in the following question. To what extent should professional and scientific standards and norms be left to the professions themselves? To what extent should the public have a role in

establishing the ethical standards for the professions? This
is, of course, a hotly-debated point these days in great part
because the medical and scientific communities often resist and
resent attempts on the part of the public to lay down standards
to be applied within the profession. It is often argued that
the general public cannot understand the difficult scientific
issues at stake and that, in any case, the scientific and medical professions composed of honorable human beings, can take
care of its own ethical issues. There is a great deal of merit
to that argument. Both in medicine and science, there has been
a tradition of ethics and virtue which few other professions
historically have been able to approach. At the same time, it
has also been well-recognized in our culture that no group ought
to be left entirely to be its own judge in matters which affect
the public welfare. It is not necessarily that any one group
is morally more suspect than any other. Rather, the possibility of self-interest makes it imperative that others also take
a look at what is going on within professions and have some say
about that. Moreover, it is an important insight of our time
that what goes on within the scientific and medical professions
can have enormous ramifications for the whole of human society.
As mentioned earlier, if the implications of the work standards
of a profession will affect others, then those others--in this
case the general public--should have some say about what happens to them.

SUMMARY

By and large, I have stressed questions, only hinting here
and there at possible answers. Certainly our society as a whole
has no answers to those questions. One finds heated debate and
difference on practically every problem I have raised here.
Many of us as individuals have come to some answers and solutions, and we are prepared to live with them. However, it seems
to me that we ought to be basically muddled and upset by these
questions. I think there are no ready and obvious answers in
sight. The Western philosophical and religious tradition provides, I believe, many basic insights and both actual and potential general guidelines. But those guidelines must always
be re-interpreted in the light of new knowledge and of changing
social conditions, changing definitions, and changing medical
knowledge and practice.

Life does not stand still, and it is surely not standing
still now. Every new generation must go back to its sources,
reexamine its present problems in their light, and decide what
it ought to do. Our generation is perhaps faced with the most
critical decisions human beings have ever had to make. We may

bemoan the fact that we have so much power, or we may be thankful that we have it. But we must ruefully admit that there is nothing more hazardous, confusing, and in many circumstances noxious, than limited power. We can do some things, but we cannot do all things.

The central ethical problem is to determine how—as human beings have always had to do—to live with our finitude. That the power of modern biomedicine seems to tantalize us with the possibility of transcending that finitude is perhaps the greatest seduction of all. But we are not going to transcend it, and our power will never be complete. Indeed, we might well ask to what extent we even want greater power, or whether we have too much power already. Of course there is little likelihood—should we decide that we have too much power as it is—that time can be reversed, much less the course of human culture.

We are in for a period of long struggling and agony, and we may not be able to come out of it. In the meantime, we are responsible to do what we can, to use our power wisely, to admit its limitations, and to find ways as individuals and as societies to live with the perennial tension between strength and weakness.

THE ETHICAL CONSTRAINTS OF DECISION-MAKING

David F. Allen
Yale University Divinity School

Dr. Callahan has presented an excellent description of the
major dilemmas facing human society as a result of the recent
advances in biomedical technology. I personally appreciated
the clarity and precision of the questions used to demonstrate
the ethical issues. In the biomedical arena there are many
issues, conflicting opinions, and few solutions. However, the
first step toward a good solution of any dilemma is to ask appropriate questions.

Dr. Callahan emphasizes that the biomedical revolution ac-
centuates the dilemma of increased, though limited, power
which results from new but limited knowledge. He warns that
although increased knowledge and power have the potential for
good, they may be used for evil as well. According to
Callahan, "Life never comes neat and complete; it always
seems to present us with the perverse paradox...that the more
struggle we have, the more problems we create, the more in
danger we are of being possessed by that power. The more good
we can do, the more harm we can also do. This is what makes
life both enticing and at the same time, enormously frustrating. We are finite beings, and our power is finite as well.
Nonetheless, we do have power, and the question is how are we
to use it wisely, particularly in light of the new possibilities in biomedical technology."

DAVID ALLEN was trained in medicine at St. Andrews Univer-
sity, Scotland and Guys Hospital, London. He completed his
psychiatry residency at the Harvard Psychiatric Service of the
Boston City Hospital where he was Chief Resident in Psychiatry.
Dr. Allen was awarded a national Joseph P. Kennedy, Jr., Found-
ation Fellowship in Medical Ethics at Harvard University in
1973-74. He was then appointed Chief Inspector for Mental
Health of the State of Massachusetts. Presently Dr. Allen is
Associate professor in Psychiatry and Medical Ethics at the
Yale Divinity School. He has published articles on the ethical
responsibility of the physician and co-edited a book on TRENDS
IN MENTAL HEALTH EVALUATION.

What, then, are we to do? What is the controlling factor in the use of our new but limited power? Is it the technology? No, it is man himself. He can use the power of the new biomedical technologies for good or for evil. Therefore, in any discussion of the ethical issues, we must address the nature of man, his ambitions, beliefs, and values.

In speaking to the nature of man, Callahan says, "Human beings, as we know, have always been a problem to themselves. On one hand, mankind has been blessed with the gift of reason. On the other, that reason has led man to speculate about himself...to speculate both about his glories and possibilities, and also about his sins and folly."

On the subject of values, Callahan says that traditional values inherent in Judeo-Christian and Western ethics are adequate. According to Callahan, "Such traditions have stressed individual rights and human dignity, the sanctity of life and the need to protect human life, and the need to care for people in times of illness and distress."

But Callahan goes on to emphasize that the real problem facing modern society is how to interpret these values in light of the new issues. For example, how do we achieve the balance between protecting human life through medical treatment and artificially extending a person's life span beyond all sense and reason?

I sincerely agree that this is a problem. However, it would be wrong to assume that this difficulty in interpreting traditional values in light of new knowledge is unique to modern man. History is filled with scientific discoveries which have challenged Christians to reexamine their faith (e.g., the treatment of mental illness).

Dr. Callahan devotes the latter part of his paper to a discussion of ethical issues involved in the decision-making process. I agree that the decision-making process is the central issue of all the dilemmas raised by the advances of biomedical technology. Yet Dr. Callahan cautions us against reducing all ethical questions to "questions of who should make the decision." He says our first task is always to seek to clarify the central ethical issues at stake in the decision-making process. My own exposure to ethical conflicts in psychosurgery, mental retardation, and the evaluation of mental health programs has confirmed this orientation. Therefore, I intend to devote the remainder of my response to personal concerns about the decision-making process.

We must consider at least five major areas in defining the ethical constraints of the decision-making process. These are: the facts; our theological basis; our moral reasoning; our loyalties; and implementation.

FACTS

We must commit ourselves to whole-heartedly seek the facts. Good ethics begins with good data! Roderick Firth's Ideal Observer theory offers a model for the collection of data which I have found helpful.[1] Firth's Ideal Observer must be omniscient, omnipercipient, disinterested, and dispassionate.

We must always strive towards omniscience through empirical research and consultation with higher authority. This entails gathering knowledge from relevant fields as well as our own. To have good data we many need to consult with ethicists, philosophers, lawyers, politicians, ministers of religion, etc. This is most important because in a democratic society no one person or group should have a monopoly on biomedical decisions affecting the lives of many persons. In deciding whether to use psychosurgery on a violent person, for example, who can claim to be an expert in the social, cultural, psychological, medical, and criminal determinants of violence? The facts must be collected from all relevant disciplines, including the family and community.

In seeking to be omnipercipient, we must try to understand how the decision affects others...especially the interested parties. For example, in establishing a human rights committee to protect the rights of retarded persons, it is ethically mandatory that retarded persons serve on the committee. Or, in advising a pregnant woman about prenatal diagnosis, she should, if possible, have the opportunity to talk with another person who has undergone the same procedure. The basic dynamic here is *empathic caring*, which results from identifying with another and treating him as you yourself would want to be treated. In essence, it is the Principle of Reciprocity, or the Golden Rule.

The Ideal Observer must also be disinterestedly-interested and dispassionately-passionate. But how can we, in the biomedical field, ever deny particular interest or passions in order to be truly interested or passionate toward the whole?

I believe that a relative state of disinterestedness or dispassionateness can result from involving a multidisciplinary committee in the decision-making process. Particular interests or passions are diluted and balanced by those who hold opposing views or perspectives. One of the best examples of such a process is the legal jury system. Similarly, ethical dilemmas in biomedical areas may be resolved through the establishment of a medical jury system. Certainly, the power to affect lives is just as great, so the way the decision is reached should be as unbiased. For example, is it any more serious for the courts to sentence a man to life imprisonment than for medicine to irreversibly change a man's behavior by implanting electrodes in his brain? Judicial decisions involve a legal jury, while medical decisions are often made at the discretion of one, or perhaps two, medical men who already share a purely medical bias.

In my opinion, the multidisciplinary committee offers the client a broad-based societal advocacy. This not only supplies a broader basis for information, but can also be very supportive to the client and his family during the very stressful time. It engenders an atmosphere of trust and openness within the doctor-patient relationship, which is conducive to more just and better quality medical care.

But the facts alone are never enough. Facts must be related to an ethical framework in order to actually serve the best interests of mankind. Silvano Arieti says that facts devoid of the human ethical perspective force man to become an animal without freedom and dignity.[2] Surely the scientific rigor of medical technology in Nazi Germany underscores the dangers of factual pursuits without an adequate human/ethical base.

THEOLOGICAL BASE

Central in our approach to the ethical issues in biomedical technology is our belief concerning the nature and meaning of man. Theological assumptions influence our ethical decions in subtle but powerful ways.

Those who adhere to the Judeo-Christian ethic believe man is created in the image of God. This *imago Dei* is the basis of every person's God-given personhood, dignity, and human rights. In addition, God gave man dominion over the earth and charged him to act responsibly toward his fellow men, other creatures, and his environment.

However, man's rebellion and refusal to accept this God-given responsibility severed the close relationship with his Creator which man once enjoyed. This broken vertical relationship between man and God affected the horizontal relationships of man and man, man and environment, and man and

54

himself. Stamped forever with the image of God, man continues to strive to do good, but the scarring effects of sin are ever-present. Thus his power to do good is matched with his intention to do evil; his desire to love is contradicted by his affinity to hate. Such ambivalence is so aptly described by the Apostle Paul when he says, "For the good that I wish, I do not do; but I practice the very evil that I do not wish."[3]

However, we also believe that God in Jesus Christ through his sacrifical death on the cross and the power of his resurrection has brought healing to the relationship, reuniting man and his Creator. This has reestablished a strong ethical base involving love, forgiveness, and righteousness in the treatment of others.

The proof of this ethical base is not some ecstatic mystical trip or a well thought theology, but a deep sense of love and worship for the Creator. It is expressed in an increased feeling of responsibility and commitment to our fellow man.

The corollary of all this is that all men, whether retarded, mentally ill, or elderly, must be treated with respect and dignity, as persons created in the image of God. Everybody is somebody, and our brother's destiny is directly related to our own. Therefore, we must tread cautiously, wisely and with care when making decisions in biomedical technology affecting the lives of other people.

MORAL REASONING

The moral reasoning involved in any decision-making process always has a profound influence on the way persons will be affected by the decision. Therefore, we must carefully examine the types of moral reasoning operating in our biomedical decisions. Though there are many different types of moral reasoning, in my personal experience two stand out in the field of biomedical technology: utilitarianism and equality. Although there are many variants, this approach can be summed up as the greatest good for the greatest number. Though utilitarianism is often appealing, it offers nothing for those who are not part of the greatest number. This may help to explain the atrocities inflicted by biomedical technology upon unfortunate minorities--the retarded, the poor, the racially outcast. The inhumane Tuskegee syphilitic study performed on unknowing Black men, the injection of lethal hepatitis virus into retarded residents at Willowbrook, and the negligent care of many powerless elderly persons in state facilities, show what can happen to those minorities considered expendable for the greater good.

Utilitarianism assumes that only life of a certain quality is worthwhile. Personhood is defined on the basis of whether one's utility outweighs one's disutility. Whenever the utility, disutility balance is upset, one's value as a person becomes less. Thus only the powerful are strong, and they are strong only as long as they have power.

This is evil and must be condemned. It is antithetical to the Judeo-Christian ethic, personified in the person of Jesus Christ, who was rich, yet for our sakes became poor; who, though he existed in the form of God..., emptied himself, taking the form of a bondservant;[4] who was willing to give up his power-bases, and share it with those who had none.

In contrast, the equal value view of life assumes that every body counts. Personhood is a gift of God not dependent on one' utility/disutility ratio. Persons have utility, but utility should never define personhood. Thus the minority, the retarde the physically handicapped, the elderly are persons first, deserving respect and dignity regardless of their usefulness to society. Justice demands equal consideration for each person's claim, regardless of the man or his situation. To make this a reality, we must share our power base with those who have none. The age-old test of involvement remains the Golden Rule; are we willing to ensure for the powerless that which we want for ourselves, thus making advantages for the most disadvantaged?

As our power increases so does our responsibility. We now have the potential through selective euthanasia and prenatal diagnosis to eliminate many disabled persons from society. Is this what we desire? It seems so easy, the best thing for the greatest number. Yet in reality, the disabled, the weaker members are an integral part of human society. They are a part of us, and we are a part of them. The way we treat them is a direct reflection of the humaneness and quality of life we espouse. Thus any society which denies justice to its weak, which eliminates those with no power base, destroys the roots of compassion, justice, and love. Then only the powerful are safe, and they are safe only as long as they have power!

Therefore we must seriously examine our moral reasoning when we make major biomedical decisions. Even ethical restraints such as informed consent and review committees will not effect justice for those without power and branded useless by a utilitarian society.

LOYALTIES

Our loyalties dictate the ultimate ends we serve in making

ethical decisions. ("For where your treasure is, there will your heart be also"[5]). When faced with decisions, we must ask ourselves, "To whom and to what are we responsible?" We must examine both our ends and our means.

For the Christian the true end of life is to glorify God and experience the meaning of our God-given personhood. All else must be a means to that end. Biomedical technology is only a tool to effect that end. Whenever it becomes an end in itself, persons are destroyed. This is the ever-present danger, which can turn a force for creative good into a destructive evil.

Coming at a time when values are in crisis, families are disintegrating, the economy is depressed, the national spirit is all but crushed, it is even more difficult to maintain a Christian perspective of ends and means. Feeling hopeless and on the point of despair, society's tendency is to see technological achievement as a means of salvation.

Christians must be challenged, in contrast, to rededicate themselves to God through repentance and a renewed commitment to serve our fellow man. Only then will we retain a true perspective of ends and means. Only then will we witness to the true perspective of a man's achievements, seeing biomedical technology as an excellent means to glorify God and enhance the dignity of our fellow man.

IMPLEMENTATION

In the area of implementation, "where the action is," we must tread cautiously with wisdom, understanding, and openness to new information, methods, and other input.

We must ask ourselves questions such as: How will this decision affect the person involved? Would I object to the procedure being applied to myself, my children, or others close to me? Can I universalize my actions (i.e., what would happen if everyone acted as I do?) What is my basic motivation for making this decision?

Action and motivation are inseparable, and the requirement for both is justice, mercy and humility before God.[6] Translated into the area of biomedical decision-making, we must treat all persons fairly by obtaining proper informed consent and providing the maximum level of professional competence. Our actions should be merciful and compassionate, treating others as we would want to be treated, sharing our power base with those who have none. And always we must seek for God's guidance.

57

Because the dilemmas are many, the solutions few, and our knowledge limited, we have ample cause to be humble. We should always be willing to admit failure or seek further consultation from higher authority. We must search continually for better answers, technologies, and methods to implement principles of justice and peace for our fellow man. Our aim is not only to be successful, but above all, to be faithful to the task to which we are called.

FOOTNOTES

[1]Roderick Firth, "Ethical Absolutism and the Ideal Observer," *Philosophy and Phenomenological Research*, Vol. XII, No. 3, (March, 1952).

[2]Silvano Arieti, "Psychiatric Controversy: Man's Ethical Dimension," *American Journal of Psychiatry*, 132: 1, (Jan., 1975).

[3]Romans 7:19.

[4]Phil. 2:6-7.

[5]Matthew 6:21.

[6]Micah 6:8.

SERVANTHOOD AND THE EXERCISE OF POWER

Richard L. Spencer
Trinity Presbyterian Church, Pasadena

I am especially pleased to find in Dr. Callahan's paper a
thoughtful discussion of two frequently-overlooked issues that
are indispensable to serious ethical reflection on technology.
The first is the paradoxical character of human power. Efforts
to extend it inevitably lead to unintended limitations of human freedom. The second is the problematical character of
decision-making where the use of power is at stake. I will
comment on these further.

CRITICISMS

There are three specific criticisms that I would raise concerning the content of the paper:

(1) In his discussion of development in applied genetics,
Dr. Callahan suggests that people can now control the quantity
of children they bear and will increasingly be able to control
the quality of their children. One must assume that biological
quality alone is meant here, since parents have always had considerable influence on their offspring. I mention this mainly
because much of the literature fails to make this distinction.
I consider this failure to be an expression of a biological bias
on the part of the writers. While I do not necessarily ascribe
the same bias to Dr. Callahan, I do believe that greater care
should be taken to emphasize the nonbiological qualities in human life which are ultimately of greater value.

(2) Dr. Callahan states that the new powers we have "to
shape, manipulate and modify human behavior" have as their essence the bypassing of free choice and "finding ways to get

RICHARD L. SPENCER is Senior Pastor at Trinity Presbyterian
Church in Pasadena. He holds the B.D. and Ph.D. degrees from
Princeton Theological Seminary, the latter degree in ethics.
He also received an M.A. in clinical psychology from the Grad-
uate School of Psychology, Fuller Theological Seminary and has
served Fuller Seminary as Adjunct Professor of Ethics.

people to behave." It is not at all clear that behavior modification techniques necessarily bypass free choice, even when they are employed by parents in guiding the behavior of their children. The introduction of a purposive design into patterns of reinforcement may reduce confusion and hostility, and may in fact nurture the kinds of behaviors that are the precondition of mature, free choice. Anyone with experience in behavior modification can tell story after story of parents and their children acquiring far greater freedom to do the things that matter most to them once the minimum conditions of cooperative family behavior have been achieved. Current practice in psychiatry and clinical psychology also includes growing use of self-control techniques, such as biofeedback and behavioral self-modification.[1] Many who have benefited from them report that both the range and efficaciousness of their free choice have been measurably increased.

Behavior modification, like any other technology, can be used inappropriately. It is especially subject to abuse where counter-control measures are restricted, as in a prison. Knowledgeable and morally sensitive evaluation of its use is always appropriate. But in accusing the behavior modifiers of reducing human freedom, one must be careful to distinguish between the "freedom" of mere capriciousness and the "freedom" of lawful and orderly pursuit of human purposes.

(3) It has been suggested to us that if people could be kept alive indefinitely and with aging, "then we would have total power." I wonder. Human vitality in its most general sense seems intimately bound up with the rhythms of birth, life and death. The recent science-fiction film "Zardoz" portrays the physical and spiritual sterility of a future society in which perpetual youth is the norm and aging a form of punishment. Its members secretly longed for death, but they were powerless to achieve it. While there was much about the film that I did not like, its central message contained an element of truth. That message might be stated, in part at least, by the title of Dr. Callahan's recent book: THE TYRANNY OF SURVIVAL.[2] The biblically-oriented Christian must certainly find the effort to extend this bodily life ironic if not grimly amusing. Biblical faith suggests a rather different notion of "total power."

THE BIOLOGICAL AND THE HISTORICAL

My principal response is to suggest the importance of a theme which is present in the paper by implication only. Both the paradox of power and the problematical character of decisions regarding its use point to this theme. Taking them

60

seriously requires us to see that human biological life, even if perfected to a degree that is beyond our imagination, is merely the necessary but not the sufficient condition of that which is most essentially human: history.

To be historical, man must have a horizon and a promise to live by. Biological functioning is not enough, even at the most optimal levels imaginable. An appreciation of the historical quality of our common life is essential to an adequate understanding of human power, responsibility, and decision-making. Human beings have a capacity for worldly freedom and self-transcendence that distinguishes them from the rest of the natural order. It is a capacity which, corrupted by sin and touched by God's grace, finds its expression in the contemporary condition of mankind one marked by paradox and ambiguity. Biblical faith holds that this paradox and ambiguity are inevitable within the present order of things. The human capacity for good and evil is, if I understand the Book of Revelation correctly, increasingly powerful in its benevolent and malevolent effects. Even the claim to represent or to have achieved the greatest good may itself mask or rationalize outright evil. The problem of Christian moral discernment and service of the good is, within the theology of history, a formidable one.

I would like to make three particular points in this regard. First, God's revelation to us in Christ is the central and controlling warrant for Christian moral commitment. The character of that revelation, that historical appearance, must instruct us concerning the appropriate Christian orientation to the world of values and moral choice--especially where the exercise of power is at stake. Its character is beautifully summarized in Stauffer's NEW TESTAMENT THEOLOGY. Stauffer finds two competing principles in the New Testament. The operating principle of the rebellious and sinful world, Stauffer says, is *superbia*: pride and self glorification. When God came to that rebellious world in Jesus Christ, He came not on its terms but on His He came by way of the "soft underbelly" of the world--by way of *humilitas*: humility and servanthood.

The second chapter of Paul's letter to the Philippians describes "the mind which we have in Christ Jesus" very much in these terms. Jesus Christ has confronted human power and pretense precisely at the place of its greatest vulnerability. He is crucified because He threatens the established system of worldly power and its *superbia*. As Stauffer has written:

The glory with which God has adorned His creation has become His temptation. Consequently, the freedom

61

which God gave Him works His ruin. The creature
means to become something, something without God,
something like God--and, if need be, in spite of
God. So the prime motive of demonic activity is
self-glorification. From now on the struggle between one glory and another, between the *gloria dei*
and the *gloria mundi* is the dramatic theme of all
history, even of the history of the Church.... So
the NT opposes *superbia* in all its forms, among
believers as well as unbelievers.[3]

The Christian faced with moral choices regarding the uses
of human power is called upon to follow the way of *humilitas*.
His orientation to power is servanthood, i.e., seeking the
good of the other without consideration for self-glorification
or the increase of personal, worldly power. Influence and
technique are means in service of the compassionate, personal
ends appropriate to Christian love.

Our increasing power over human biology and behavior confront us with new opportunities for service and new temptations toward utopian self-glorification of the race. History
is replete with the bitter, unintended consequences of utopian
causes. The Christian may wish to survey these powers and
their use with a mind informed by biblical wisdom and some
historical sophistication regarding the temptations of utopianism. It is my conviction that the exercise of human powers
done wisely and in the spirit of *humilitas* will usually lead
to consequences, both intended and unintended, that are benevolent. Even here, though, the Christian remains aware of the
necessary historical ambiguity of deeds that grow from even
the best of motives. The taint of sin and guilt are never ab-
sent from human projects this side of the full fruition of
Christ's kingdom. However, this recognition does not need to
rob the Christian of the fullest joy that can flow from life-
giving service. Conscious participation in the dialectic of
sin and grace provides some protection against the seductive
appeals of utopian pride. The Christian seeks to do good while
seeking to avoid doing evil in the name of good.

Second, the crucible of human development in history is not
biology but relationship. The web of connection and interactions between intimate family members, on the one hand, and the
larger human community, on the other, gives shape to human character. I believe that this pattern of relationship extends
all the way back to the womb. The subtle interactions between
mother and fetus may prove to be immensely important and quite
simply irreplacable by any technological achievement. I am
thinking here of *in vitro* fertilization and of the suggestion

62

that human gestation may one day take place in a completely artificial environment (the so-called "test-tube baby").

If human relationship is essential to man's relationship with God, fundamental interference with the process of natural gestation might have consequences unacceptable to the Christian. The historically sensitive Christian will seek to preserve the priority of relationship over biology where the two may come into conflict; he will do so as an expression of a fundamental value-judgment.

Third, all of us are shaped by our own finite and localized exposure to history. Our perceptions of truth and our sense of what is plausible are, as the sociologists of knowledge have shown us, profoundly conditioned by our most characteristic social experiences.[4] No human being is free of particular perceptions, interests and values that shape his or her response to the situations of choice and action. Informed honesty requires the open acknowledgement of our ethnocentricity, our class-bound consciousness, and the special interests embodied in our values. The universal character of the divine imperative which is known in Jesus Christ does not relieve the Christian of this particularity nor of responsibility to come to terms with it. Christ's agapeic imperative puts that particularity under divine pressure, partly by pressing the Christian into community with others who embody a somewhat different historical experience. In that encounter, especially where the exercise of power is at stake, the skewed character of our consciousness and special interests both become quite apparent. The Christian who follows the spirit of *humilitas* may find himself called to repent of his tendency to universalize and idealize his own narrow perspective. Class values and social prejudices may be upset in the process, but the way of *humilitas* should find such congenial. Indeed, the Christian should seek out that community in which judgments regarding the use of power are subject to examination by those who have a different stake in its use, especially when persons are themselves the direct objects of the power in question.

Participation in the meritocracy of higher education and professional life is itself a form of narrow existence leading to biases, perceptions and partial values. Those of us who share in it need to admit our narrowness and confront our wider world with open eyes and minds. We also need to recognize the dangers implicit in any policy-setting that is done by those sharing a single set of social values and experiences. We should, instead, seek to include those shaped by differing social experiences in the process of setting policy and establishing ethical guidelines. Saying these things is not to deny that our theological warrants have universal validity. It is merely to assert that our employment of them tends to reflect our historical particularity more than we are likely to guess. We must especially guard against the abuse of power in areas of human life especially sensitive to value judgments by precisely those having the best intentions.

FOOTNOTES

[1] A very good example of the behavioral approach to self-modification is Michael J. Mahoney and Carl E. Thoreson, *Self-Control: Power to the Person* (Monterey, Calif.: Brooks/Cole, 1974).

[2] Daniel Callahan, *The Tyranny of Survival: And Other Pathologies of Civilized Life* (New York: MacMillan & Co., 1973).

[3] Ethelbert Stauffer, *New Testament Theology*, trans. John Marsh (London: SCM Press, 1963).

[4] An especially instructive volume in this regard is Peter Berger and Thomas Luckmann, *The Social Construction of Reality* (Garden City, N.Y.: Doubleday, 1966).

PART III
THEOLOGICAL PERSPECTIVES

British neuroscientist Donald Mac Kay examines the biblical principles upon which he feels assessment of human engineering needs to be based. Dr. Mac Kay begins by pointing out the need to distinguish biblical concepts of nature from Greek concepts, which have often been uncritically mixed into Christianity. He argues that the biblical view of nature encourages the search for truth and the general activity of science. Mac Kay then shows that the proper position of human beings is as God's fellow-worker. He argues that the biblical position is one of integral complementarity between dependence upon God and moral responsibility for our choices. Two of the limitations which affect human capacity to understand and carry out God's normative will are sinfulness and finiteness. Mac Kay distinguishes between these concepts and shows that although choices may be made with all proper intent, finite understanding of all the possible ramifications may bring about bad consequences.

Mac Kay then considers several biblical doctrines and shows how they relate to the challenges posed by biomedical and behavioral science. The warning is that there are limits to the extent of improvability; human engineering can never correct the sinfulness of humanity. At the same time, Christian responsiblity includes the need to make efforts to lessen the effects of that sinfulness. Mac Kay goes on to explore the meanings of compassion and ambition with regard to obligation to seek means of reducing suffering. He suggests that much of the uneasiness over human engineering research centers on the perception of it as dehumanizing. Distinctions between manipulation and dialogue are made, and Mac Kay argues that manipulation in an experimental sense may not always be unethical. The real issue is the nature of the relationship. There must be potential or real answerability, depending upon the situation. In a fascinating section on personality changes, Mac Kay wrestles with the issue of individuality and the nature of the soul in the face of permanent personality changes that a procedure such as psychosurgery might bring. Finally, Dr. Mac Kay poses eight biblical check-points which Christians can use in assessing the appropriateness of human engineering.

In response, biochemist Robert Herrmann stresses the need for adequate data that allows full understanding of choice implications. In the area of genetics he indicates the need to know much more about pleiotropic effects and to consider the ethical implications of recombinant DNA (instead of focusing purely on the biological hazards). Finally Dr. Herrmann argues that the existence of risk must not be allowed to push us to a protected, papier-mache world in which true dominion is not exercised as God intended it to be.

Theologian James Olthuis asserts that technology is not neutral. Rather, any technology needs to be evaluated by established normative standards. A given technology may be good or bad, but conception of it as neutral simply contributes to our regard of it as an ethically self-contained system not subject to external norms. It becomes good because it is, unless such normative evaluation is conducted. Olthuis suggests that the Scriptures not only teach the purpose of human beings, but also their essential identity as servants of God. Techniques which distort or destroy the capacity of human beings to responsibly serve God are judged wrong. Dr. Olthuis then says that the love principle pointed out by Mac Kay is right, but insufficient. What is needed are concrete specifications of love if one is to make consistent ethical decisions. He argues that the distinction Mac Kay makes between the creative and normative will of God should not be made. Olthuis then suggests some more specific guidelines based on the creation norms. Finally, he briefly illustrates the application of the norm network that he has suggested to issues of artificial insemination, *in vitro* fertilization and cloning.

BIBLICAL PERSPECTIVES ON HUMAN ENGINEERING

Donald M. MacKay
University of Keele, England

What new obligations are laid upon us by the prospect that
human behavior, and the very constitution of human beings, may
be shaped by scientific means? In principle, both behavior
control and human stock-breeding are as old as civilization
itself; but in our generation, for the first time, we are having to face possibilities beyond anything considered in traditional ethics. We want to do good, if any good can be done
with our new knowledge. At the same time we are unhappily
aware of the ethical risks inherent in the manipulation of
human beings, even if our intentions are the best. Torn by
the claims of compassion and of respect for the individual,
men of goodwill look to one another for guidance and find it
hard to see a clear way forward.

In relation to these issues, what difference should it make
to be committed to the biblical Christian faith? There is no
suggestion that the Bible contains ready-made answers, still
less that Christians already know what the answers should be.
Indeed, Christians feel the tensions no less painfully than
others.

DONALD M. MAC KAY is professor and chairman of the Department of Communication at the University of Keele, England. He
has done extensive research and writing in analogue computing
and brain physiology. Dr. MacKay has published over 100 scientific articles and authored THE CLOCKWORK IMAGE, CHRISTIANITY
IN A MECHANISTIC UNIVERSE, and FREEDOM OF ACTION IN A MECHANIS-
TIC UNIVERSE. He is co-editor of EXPERIMENTAL BRAIN RESEARCH
and HANDBOOK OF SENSORY PHYSIOLOGY. He is a member of the
Physiological Society and Institute of Physics in England. He
has delivered the Edington Lecture at Cambridge, the Herter
Lecture at John Hopkins University and the Forester Lecture
at U.C. Berkeley. Along with frequent appearances on British
television, he has dialogued with B.F. Skinner on William F.
Buckley's television program, Firing Line, in the 1971 program
entitled "Case Against Freedom."

The danger in all such inquiries is that we do not start far enough back; we do not set our problems in a wide enough context to avoid inconsistencies in our thinking. My plan in this paper is therefore to begin at ground level by considering first how the whole approach of science and technology fits into a biblical view of the world. Once our theological lines are clear, we can consider more specific issues raised by the notion of "human engineering." It should be obvious that my aim will not be to offer definitive conclusions, but to clarify the biblical principles upon which Christians will have to base their assessment of this challenging situation.

BIBLICAL OR PAGAN?

For anyone, Christian or otherwise, who wants to understand how biblical Christianity relates to science and technology, the first priority is to distinguish between genuinely biblical concepts of nature and man and a variety of pagan ideas that are commonly confused with them. For many people in our day, as in previous ages, the typically 'religious' attitude toward natural science is supposed to be that embodied in the ancient Greek legend of Prometheus, who stole the sacred fire. Nature is thought of as semi-divine and has her secrets. The gods would like to keep some of these to themselves, and jealously resent any advances in man's knowledge of them. Science is thus an irreverent and dangerous pursuit in which sinful man aspires unto the place of God. If disaster results from attempts to apply man's scientific knowledge, this is what he deserves for prying into the sacred mysteries of the Creator.

Now it cannot be denied that if your idea of God (or the gods) were that of the ancient Greeks, indeed of almost any pagan religion, all this would make good sense. To some people it might seem to represent the proper humility of man before the majesty of his Maker. But is it in fact biblical? I think not. The Bible has no time for human pride, but its teaching about the natural world is precisely the reverse of the pagan at crucial points.

The Bible sets man in perspective as a creature of God, a part of the vast created order that owes its continuance in being to the divine upholding power. Unlike the rest of the natural world known to us, however, human beings have powers of foresight, planning and action that make us specially responsible in the eyes of our Creator. With these powers, according to the Bible, goes a special obligation toward the Creator. Men are commanded, not merely permitted, to "subdue the earth" (Gen. 1:28). This is not to be done, indeed, in a spirit of arrogant independence, but as the stewards of God's creation.

68

Human beings are answerable to Him for the effectiveness with which they have fulfilled His mandate. Our overriding priority from the biblical standpoint is to love God and our neighbor. All human exploitation of natural laws and resources must be an expression of this love, and of nothing else, if it is to be acceptable.

The Christian ethos is in complete contrast to the pagan caricature with which it is so often confused. In place of craven fear that haunts the unwelcome interloper, we are meant to enjoy the peaceful confidence of a servant-son at home in his Father's creation. We know that we are on our Father's business no less when investigating His handiwork than when engaged in formal acts of worship. In place of jealously se-cretive gods we have One whose very nature is Truth, and Light, Himself the giver of all that is true, who rejoices when any of His truth is brought to the light and obeyed in humility (e.g., I John, *passim*).

The Bible encourages man to roam the domain of the natural world in responsible freedom, showing all of it the respect due to his Father's creation, but none of it the superstitious reverence that would deny its status as a created thing like himself. As Professor Hooykaas has put it,

> The Bible knows nothing of 'Nature' but knows only 'creatures', who are absolutely dependent for their origin and existence on the will of God. Consequent-ly, the natural world is admired as God's work and as evidence of its creator, but it is never adored. Nature can arouse in man a feeling of awe, but this is conquered by the knowledge that man is God's fel-low-worker who shares with Him the rule of the fel-low-creatures, the dominion over the fish of the sea, and over the fowl of the air, and over the cat-tle, and over all the earth... Thus, in total con-tradiction to pagan religion, nature is not a deity to be feared and worshipped, but a work of God to be admired, studied and managed. In the Bible God and nature are no longer both opposed to man, but God and man together confront nature.[1]

GOD'S FELLOW-WORKER

The biblical concept of man as God's fellow-worker is not without its logical difficulties. If God is almighty, why does He need our help? If He has willed things thus and so, how can any action of ours improve upon His presumably perfect will?

The answer sometimes offered is that God voluntarily sets limits to His power, and leaves us room to supplement His action. According to this model God does so much, and man's part is to do the rest. I will not go into the further theological difficulties that are raised by such an answer. All I would say now is that it is emphatically not the answer offered in the Bible itself.

For the biblical writers there is no question of any such partition of action between God and man. "Work out your own salvation with fear and trembling, for it is God who is at work in you to will and to do of His good purpose" (Phil. 2:12-13). This injunction was given to New Testament Christians. The Old Testament is just as clear that in one sense at least all men, whether they love God or hate Him, are giving expression to the creative purposes of God by their choices and actions. "You thought evil against me," says Joseph to his brothers, "but God meant it unto good..." (Gen. 50:20). God is the immediate giver of being to all that is and all that moves, the wicked as well as the good. In a profound sense the whole drama of creation unfolds according to His "determinate counsel and foreknowledge" (Acts 2:23).

It would thus be a logical blunder to interpret human responsibility in the Bible as something that takes over where God leaves off. The Bible clearly represents us as both wholly dependent on God for every event of our existence, and wholly answerable to Him for the responses we make. The slogan: "Work as though all depended on you; pray as though all depended on God" may be somewhat oversimplified, but it comes far closer to expressing the spirit of biblical realism than any attempt to parcel out zones of responsibility between God and man.

This is not the place to spell out the logical fallacy in attempts to make a contradiction out of these complementary emphases.[2] Suffice it here to say that if we pay attention to the difference in logical standpoint between talk about a creator (any creator) and talk about his creatures, it becomes clear that the agency of the creator is not an alternative but a necessary condition of the agency of his creatures. This does not make the creator morally answerable for the actions of his creatures (it would make no sense to hold Shakespeare guilty as an accessory to the murder committed by Macbeth!). Nor does it abolish the moral responsibility of the creatures for the exercise of their created capacities. But if, as the Bible declares, our Creator is One to whom it makes sense to pray, then it makes abundantly realistic sense for us to acknowledge in prayer our total dependence on Him. Simultaneously, as agents within His created drama, we recognize also our full responsiblity for our action or inaction, and the

70

logical absurdity of shrugging any of it off on to him.

To put it in another way, it is essential to distinguish between two quite distinct meanings of the will of God. One, denoting what we might call His *creative* will, is what is expressed in the Genesis phrase "Let there be..." Any idea of our going contrary to God's creative will is strictly meaningless, since apart from His creative word nothing happens. "He upholds (gives continued being to) all things by the word of His power" (Heb. 1:3).

The other concept we might term God's *normative* will. This is what is expressed, for example, in the Ten Commandments or the Sermon on the Mount, in the words "Thou shalt..." The idea of our going contrary to God's normative will is, alas, far from meaningless, however wrong and unrealistic it may be. Without God's help, according to the Bible we will lack both the ability to recognize His normative will and the desire and strength to carry it through. The gift of vision and strength to do God's normative will is what the Christian knows as Grace. It is mediated through God's creative will. It is the daily experience of living in dependence on the grace of God, which unifies the complementary doctrines of divine sovereignty and human responsibility.

HOW THINGS CAN GO WRONG

As a would-be servant of his Creator, man suffers from two limitations that we must take care not to confuse. The first is his sinfulness. We are by fallen nature headstrong, rebellious, reluctant to accept wholeheartedly the "kingly rule of God." The second we may term simply his finiteness. At his best, and with the best will in the world, a man can take only a limited number of factors into account when planning to do what he believes to be God's normative will. Because both his knowledge and his foresight are limited, things can go painfully wrong in ways that would be superficial and cruel to attribute simply to human sinfulness.

This is not to deny that our sinfulness makes things worse, but to point out that the best motives afford no automatic exemption from the unforeseeabe risks of experimentation. It is both unnecessary and misleading, for example, to write down the development of the Dust Bowls simply to human greed. The most selfless humanitarians, eager to increase the supply of food for the starving of the world, might have fallen into the same ecological trap as the hapless first settlers of the Tennessee Valley. Again, the most conscientious steward of God's creation, totally devoid of any greed might have been forgiven

71

for thinking that DDT spraying was the responsible thing to do on a large scale in subduing the earth for the benefit of mankind. The temptation to ferret around for some ingredient of "sin" to blame when these things turn out disastrously must be resisted if Christians are to think biblically and realistically about their wider responsibilities. Selfish unheeding of foreseeable cost and risks is indeed inexcusable, but man at his best is only, as Pascal called him, a "thinking reed." What hindsight allows the rest of us to condemn as short-sightedness is sometimes an inescapable aspect of our being human. "Let him that thinketh he standeth take heed lest he fall" (1 Cor. 10:12).

So when we turn to consider the possibilities for good in "human engineering", it is important not to imagine that the Bible's one prerequisite for success is the elimination of sinful motives and the adoption of worthy goals. We are feeling our way to the controls of a world whose mechanism is more complex and delicately balanced than we are ever likely to comprehend. What the proverbial bull could do in a china shop is nothing compared with the havoc we could wreak by a single well-intentioned error. The biblical moral is not that we should leave well alone. All is far from well, and it may be our responsibility in God's sight to do something about it. The moral is that, if we are not to make matters disastrously worse by our meddling, we shall need a Wisdom infinitely greater than our own. If our Creator is willing to give this wisdom to those who ask in humility and sincerity, desiring only to be used by Him for good, then nothing could be more realistic than to beg it from Him "who giveth to all men liberally, and upbraideth not" (Jas. 1:5). We are likely to need the reassurance of those last three words!

THE BIBLE AND THE SCIENTIFIC ENTERPRISE

The Bible contains no worked-out 'theory of nature'; its concern is with more urgent matters. The biblical view of the natural world finds expression more or less incidentally and implicitly. Nevertheless, as we have noted, enough is said to make a strong contrast with pagan attitudes inimical to the spirit of science as we understand it. Hooykaas[3] cites a wealth of illustrations from medieval and later writers to show how the fertilizing influence of biblical thought counteracted the Greek tendencies to deify nature, overvalue speculative arguments *a priori* and undervalue manual labour, which hindered the rise of experimental science. The 16th/17th century founders of the experimental approach drew encouragement from the Bible. They regarded pure science as a reverent unfolding and reading of "God's Book of Nature." Technology was viewed as

72

a compassionate endeavour to ameliorate the lot of mankind through active obedience to what that Book revealed.

Science thus emerges in biblical perspective as an activity pleasing to God. The preacher may sound rather doleful as he spells out the scholar's mandate: "I gave my heart to seek and search out by wisdom concerning all things that are done under heaven; this sore travail hath God given to the sons of man to be exercised therewith" (Eccles. 1:13). It is, however, clear from less pessimistic passages that "seeking out the marvelous works of the Lord" is a natural response of "those who take pleasure in them" (Ps. 111:2). There is no suggestion in the Bible that any domain is closed to human inquiry. Although the Creator's thoughts are as far above human thought as the heavens are above the earth (Isaiah 55:), the regularities of the natural order are tokens of His faithfulness to us. They are to be observed and relied upon by us for our good (Gen. 8:22, 9:16). They are worthy of our investigation.

In all this the emphasis on obedience to truth is central. In science, facts reign supreme. Opinion is (officially) tenable only as far as the evidence demands. In practice both anomalous exceptions to laws and baseless flights of fancy may have to be tolerated, but everyone knows where the last word lies. Since from the biblical standpoint the data that claim the scientist's obedience are God's data, Christians have particular reason to enthusiastically support this emphasis.

In summary, then, the biblical attitude to science is well balanced. The Christian who wants to think biblically will neither deify science nor vilify it. On the one hand, he can never give absolute or overriding priority to the claims of scientific investigation over ethical or moral considerations. On the other, he will resist all efforts, however well meaning to denigrate the scientific enterprise as such, or to encourage fearful pagan attitudes towards it.

TECHNOLOGY IN BIBLICAL PERSPECTIVE

The contrast between biblical and pagan theologies of nature is at no point more decisive than where science comes to inspire technology. "What right has man to improve upon nature; aren't we beginning to usurp the prerogatives of the Creator?" Such questions are often asked rhetorically, backed by observations of the kind parodied by Flanders and Swann: "If God had meant men to fly, he would never have given us the railways."

There is of course a sober warning for all ages in the story of the Tower of Babel, where men sought to build "a tower whose top may reach unto heaven" (Gen. 11:4). What the context makes clear, however, is that their sin consisted not in the building but in the motivation for it--an arrogant desire to be independent of God. Nowhere in the Bible is technological achievement disapproved, except where it expressed human pride and vainglory. More relevant is the reiterated biblical teaching that "He that knoweth to do good and doeth it not, to him it is sin" (Jas. 4:17). From the biblical standpoint whatever needs to be done to alleviate the lot of our fellow men is a duty from which we can excuse ourselves only for good cause.

The contrary pagan notion that it is both impossible and illicit for man to compete with or improve upon nature has had a long and fascinating history from ancient times.[3] The Greek concept of the Golden Age, when men were supposed to have lived healthy and contented lives without technological aids, colored much classical and medieval thinking. The supposed divinity of nature was taken to imply that man would be claiming divine prerogatives if he attempted to copy or improve upon it. The general belief of the Middle Ages was that feats of nature could be surpassed only by magic.

The most powerful biblical arguments against this pessimistic view were advanced by Francis Bacon:

> If there be any humility towards the Creator, if there be any reverence for or disposition to magnify His works, if there be any charity for man, ... dismiss those preposterous philosophies which have led experience captive, and approach with humility and veneration to unroll the volume of creation.[4]

As Hooykaas puts it, Bacon

> blew the trumpet in the war against the sins of laziness, despair, pride, and ignorance; and he urged his contemporaries, for the sake of God and their neighbors, to re-assume the rights that God had given them and to restore that dominion over nature which God had allotted to man. His ideal was a science in the service of man, as the result of restoration of the rule of man over nature. This to him was not a purely human but a divinely inspired work: 'The beginning is from God... the Father of lights'.[5]

It is worth noting that traces of the Graeco-medieval tradition lingered even in such a champion of biblical Christianity

as C.S. Lewis, who justified his anti-technological bias by
identifying human dominion over nature with hubris, commend-
ing instead the (Stoic) 'wisdom' of "conforming the soul to
reality."[6] Significantly, perhaps, he did not adduce biblical
support for this attitude! As Hooykaas comments:

> It is true that the results of our dominion over
> nature have been unhealthy in many cases; the pow-
> erful river of modern science and technology has
> often caused disastrous inundations. But by com-
> parison the contemplative, almost medieval, vision
> that is offered as an alternative would be a stag-
> nant pool.[7]

For Bacon "the entry into the Kingdom of man, founded on the
sciences, is not very different from the entry into the Kingdom
of Heaven, whereinto none may enter save as a little child."
What must be said, however, is that Bacon's repeated warnings
against the dangers of power divorced from humility and charity
have all too often been ignored by later inheritors of the scientific tradition. Under the banner of "Scientific Humanism",
the most extravagent boasts are to be heard even today concerning the limitless power of man to control his own destiny -
free, of course, from "pre-scientific" belief in God. On this
sort of thing the Bible has terse comments (Ps. 14:1), and it
certainly deserves all that C.S. Lewis had to say so effectively against it. It does not prevent us from asking whether there
is not a properly biblical way of exercising the powers that
scientific technology puts at our disposal, so that we escape
the Bible's condemnation of pride without falling under its
condemnation of neglect.

THE "PERFECTIBILITY" OF MAN

How do these general principles apply in relation to the science of man himself? Few topics so rapidly bring out the irrational in educated people as that of human perfectibility.
On the one hand are those who take it as an affront if one questions whether man is perfectible by his own efforts. On the
other, religious people see an affront to biblical doctrine in
the suggestion that any human failings could be remedied by
psychological or physiological means. For present purposes,
I suggest we discuss the more modest question of the improvability of man. What comments does the Bible offer, directly
or by implication, on the prospects of human efforts to improve
the human race?

Fundamental to this issue is the question: What are people
for? Until we settle this, all talk of improvement is indeterminate. Conversely, to the extent that we spell out what we see

as improvements in man, we are by implication answering this question. The Bible sets out the primary answer in uncompromising terms: "Human beings exist to glorify God and to enjoy Him for ever." The words are those of the Shorter Catechism[8] but the thought is solidly biblical. Human life is fulfilled in glorifying and enjoying God at three levels of relationship: (1) the individual relationship in which each person would ideally express his love for his Creator by constant perfect and glad obedience and thankful personal communion; (2) God has made people for the enjoyment of one another in family bonds of mutual expectations and love, in and through which they are also meant to enjoy fellowship with Him; (3) the Bible also lays great stress on various corporate relationships whereby people outside the family circle can become bound to one another by shared expectations and reciprocal giving. They are able to express a corporate commitment to love and serve their God, such as in a church or a nation.

In biblical terms, then, improvement must mean enchancing men's capacities for one or all of these levels of relationship to God and other people. This subsumes, but does not exclude, the development of natural human talents, whether intellectual, aesthetic or any other. But it focuses on essential criteria of improvement which goes deeper than any merely psychometric assessment. There may be any number of techniques to improve human capacities, even for religious performances, which would miss the main point of what life is for (witness the Pharisee of Jesus' famous parable). Any program to improve the human condition which fails to measure up to the appropriate criteria at these three levels must from a biblical standpoint be seen as short-sighted or worse.

What the Bible emphasizes, and our experiences show all too well, is the extent to which individually desirable factors may interact negatively. "Geshurum waxed fat--and kicked" (Deut. 32:15). "It is better... to enter into life with one eye (or hand, or foot) than having two... to be cast into hell fire" (Mark 9:42-50). "Not many wise men after the flesh, not many mighty... are called" (I Cor. 1:26). If it was so easy for well meaning people to make Dust Bowls of their physical environment, what Dust Bowls of the human spirit might not be created by analogous psycho-biological developments? Are some of these with us already?

Examples are easy to think of. Health is good, but artificially stockbreeding or cloning the finest specimens of manhood at the cost of destroying the biblical ideal of family relationships would be health at too high a cost. Self-respect is good, but a policy that gave top priority to the development of self-esteem and independence of mind, to the detriment of unselfishness and compassion, would be disastrous. Contentment is good, but the greatest unkindness we could do to a man would be to blind him to the tragedy of living with his back towards God.

It would perhaps be a relief to conclude that the weight of biblical evidence is against our even entering such a hazardous field, but I can find no such evidence. There are a number of principles that carefully restrict the kind of interference with human beings that might be legitimate, but it cannot honestly be said that the Bible proscribes all thought about improving either the individual or the race by artificial means. Therefore, we cannot, on biblical grounds, dodge the responsibility of asking whether and under what conditions there is good to be done in God's service by these means.

This point is most clearly felt if we apply it to ourselves. We might be tempted to imagine that it would show proper humility to accept life passively and learn to live with all our defects and limitations. Are we not instructed, after all, that "godliness with contentment is great gain" (1 Tim. 6:6)? It seems clear that we need a distinction between contentment with the unalterable, and complacency with the alterable. If our lives were purely self-centered it could be argued that lingering in a defective but remediable condition is our own affair. But if our primary obligation is to serve and glorify God, and if an innocuous remedy were at hand that would increase our effectiveness to that end, complacency on our part would be clearly culpable. The question then would not be: why?, but: why not? It should pain us to offer our Master any needless imperfection if we daily "present our bodies a living sacrifice" (Rom. 12:1). Whether it does or not, we rob Him by doing nothing about it. There is no more penetrating criterion than this one: What (with all its costs and implications) will make the most acceptable possible offering to God? To apply this in the case of other people, especially non-Christians, may be more difficult in detail. In essence, I suggest, the criterion must be the same: "Will a man rob God?" (Mal. 3:8).

THE FALLENNESS OF THE NATURAL ORDER

At this point we must come to grips with one of the most crucial, yet mysterious, of biblical doctrines: The "fallenness" of the natural order, including man himself. "Cursed is the ground for man's sake" says the Creator (Gen. 3:17). The whole creation has been made "subject to frustration", and "groaneth and travaileth", says Paul in Romans 8:20, 22. The divine drama in which we find ourselves is emphatically not defined as the best of all possible worlds. Our divinely appointed task is to subdue (for legitimate purposes) a natural order infested with metaphorical thorns and thistles. Moreover we ourselves, at our best, are not free from the entail of the curse. Our highest motives are likely to be tainted and our moral perceptions blunted or distorted.

In this doctrine, then, we find both a mandate and a warning. Because ours is a world full of imperfection, our charge to improve on things as we find them through obedience to the facts of nature involves no suggestion of impiety. Christ's own example makes clear that to mitigate the effects of the Fall is a duty, not a sin. On the other hand, because we are radically flawed, the enterprise of inventing and selecting among possible improvements is bound to be more hazardous than if we were not. History regrettably shows that even official membership in the Christian church is no guarantee against being taken captive by a spirit alien to that of Christ and thinking we do God service in exercising it.

We now come to a crucial distinction obscured in much of today's fashionable theology. The tragedy of fallen man is his loss of good standing with his Creator. He is not only unfit for the relationship of a son with his Father; by nature he is not even keen on it. This is his sin. As a result, he finds himself in a world that offers frustration in a thousand forms, including his own physical and mental infirmity and and a general propensity for antisocial acts and attitudes. Man needs rescuing. The chief theme of the Bible is the great and mysterious rescue operation whereby God in Christ has reconciled us to himself, and offers us each the restoration of our standing. This new relationship, unbroken even by death, is called "eternal life."

Unfortunately in some demythologizing circles the term salvation has found itself downgraded to mean little more than health - though still invested with emotive religious overtones. The result is particularly confusing in the present context, since it seems to suggest that the Gospel of human engineering could be two alternative or even competitive ways of achieving the same end. The truth is quite otherwise. The root of the biblical concept of salvation is the restoration of our broken relationship with God (cf. Romans, passim). One of its by-products should indeed be spiritual health and wholeness. However, to be interested only in the by-product and to ignore or spurn the new relationship of which it is the fruit would be to unequivocally reject what the Gospel means by salvation.

To put it otherwise, in Christ's estimation healthy people

78

need God's salvation just as much as sick people. "Improved" human stock will miss the way of eternal life without it, just as completely as "unimproved." They may even be at greater risk of doing so (John 9:41). True, there was a notable occasion, reported by all three synoptists (Matt. 9:12; Mark 2:17; Luke 5:31), when Jesus told the Pharisees that "they that be whole need not a physician, but they that are sick". Anyone, though, who has missed the irony in that remark should read the context again.

The biblical maxim is "These ought ye to have done, and not to leave the other undone" (Matt. 23:23). Human engineering at its utopian best is no substitute for divine salvation. Conversely, to give top priority to salvation is no excuse in Christ's eyes for neglecting human health and the possibility of improving it. For Christians to stigmatize (or atheists to advocate) human engineering as an attempt by man to do what the Bible says only God can do is a theological blunder.

COMPASSION VERSUS AMBITION

There is a further distinction that may prove important. To alleviate suffering is a clear Christian duty. If, for example, brain operations for the relief of anxiety raise ethical questions, many of us would feel that the onus of proof should be on those who would abolish them. It is by no means so clearly a duty, however, to seek to change the genetic balance of a human population, or the personality of a healthy individual, in directions we think to be good. Compassion for our fellow men and their descendants is not the same as ambition for them. This is not to say that one is good and the other bad, but that the second needs more careful justification, and more ruthless scrutiny of our motives. Moreover, one may shade imperfectibility into the other.

Suppose, to take a science fiction example, that we found a human population contented with a simple life, but genetically lacking in any capacity to appreciate great music, drama or the like. No one is consciously suffering, for none of them realizes what they are missing. They may even scoff contentedly at the enthusiasm of cultural missionaries for "unintelligible gobbledygook". What right, we may ask, has anyone to interfere? But to deal rhetorically with the matter in this way is not good enough. If it were not a matter of genetic defect, but simply one of educational lack, many of us would side with the pioneers of mass education in asking the converse question: what right have we to deny them ennobling experiences we enjoy, if they could be humanely equipped to share them? From this point of view, to possess the means of enlarging people's capacities for what we regard as good, yet to make no effort to make them available, makes us at least partly responsible for their deprivation.

This of course does not settle the matter. It can be cogently questioned whether some of the pioneers of mass education reckoned adequately with the potential for corruption which they were bound to enlarge along with that for good. If they could see the results today, what would they think? It is equally far from obvious whether the contentedly uncultured of our parable would be better people, by either biblical or secular standards, if their aesthetic blindspot could be removed. What the parable should alert us to is both the ease with which crass thoughtlessness for others can masquerade as the liberal spirit of "live and let live" and, on the other hand, the dangers of having ambitions for other people.

Our science fiction case can bring out one further point. Suppose that we had no remedy for the defect in the present generation, but could readily persuade them to let us give their offspring a better aesthetic endowment. We envisage that community fifty years hence, rejoicing in capacities that they would hate to lose, and glad that someone persuaded their parents to have the necessary operation. This may give us a feeling of obligation not to let the next generation down. But there is a logical fallacy in this which it is important to spot. Suppose that we had in our science fiction world ten radically different possible genetic recipes for the next generation. In that case we must by our earlier logic envisage not just one thankful community fifty years hence, but ten different alternative ones, each perhaps thankful for different things. To which, then do we owe our obligation? The answer surely is: to none of them; for they are non-existent and not even identified until we have acted. The notion of obligation to posterity becomes logically incoherent if the action in question has to determine who posterity shall be.

This does not mean, from the biblical standpoint, that we have no obligations. It means rather that we must recognize our obligation as directly to our Creator, rather than to non-existent, potential human beings. This is logically and theologically parallel to the everyday situation in which Christian parents decide whether or not to have another child. It would be nonsensical to construe their decision in terms of obligations to the future child, since it is the existence of of that child that is at issue. Perhaps in the majority of such cases the child never will exist. On the other hand, it is both meaningful and relevantto construe it in terms of their obligation to God as fellow-workers or procreators.

The suggestion that emerges, then, is that if we were ever called to exercise the function of genetic engineer, the logic and theology applicable would be those relevant to procreation. The Bible has plenty of worked examples of what it means to exercise this function both responsibly and irresponsibly, and of the special kind of mutual responsibility that is meant to develop in due course between procreator and procreated when both survive together (e.g., Eph. 6). This might provide a useful biblical model for the corresponding relationship between the genetic engineer and the offspring of his selective action.

DEHUMANIZATION

Behind much of our instinctive uneasiness with the notion of human engineering lies the feeling that it dehumanizes its subjects. It will be important for us to try to pin down this feeling and to articulate biblical guidelines for deciding which forms of manipulation, in what circumstances, are indeed dehumanizing.

To dehumanize a fellow man, in biblical terms, is to refuse to love him as oneself. It is to "shut up one's bowels of compassion from him", to harden one's heart against him, to refuse to answer his questions or to be sensitive to what it is like to be he. It is, in short, to treat him as an object.

In order to spell out what this means, and what precisely is wrong with it, we have to go into some technicalities. Manipulation requires calculation to achieve a desired end. A precondition of successful manipulation is that the process of calculation can be kept isolated from the manipulated situation sufficiently to prevent feedback which would render the calculation out of date. In the relationship of dialogue, to the extent that the goals of each partner are reciprocally open to adjustment by the other, feedback of this kind is inevitable, and manipulation correspondingly limited.[9] The mutual interaction that takes place in dialogue (as distinct from alternate monologue) makes the partners effectively one system for purposes of mechanistic analysis and prediction. With regard to certain actions each participant becomes incalculable and logically indeterminate for all the others in dialogue, as well as for himself. It would be irrational for any of them to regard such actions as unconditionally inevitable. This, I have suggested elsewhere[10], is the ethically essential sense in which normal human beings must be recognized by one another as objectively free[11]. Conversely, a calculator who withdraws from the vulnerability of dialogue may thereby enhance his powers of manipulation by increasing his calculating capacity. He may continue to communicate, but to the extent that his own goals are invulnerable he does so purely manipulatively.[12]

We have so far tended to speak of manipulation as something ethically unsatisfactory. It usually is. There are, however, certain common situations where this is not the case. In clinical psychiatry, as in other forms of medicine and surgery, for example, we encourage the patient to give himself up to be manipulated for his own good. All of us endured purely manipulative relationships at early ages for which we are now grateful.

Perhaps the most difficult borderline case, which creates much confusion today, is the education and upbringing of children. Until recently it was taken for granted that parents had both a right and a duty to mold the value-system and character of their children in accordance with the best they knew. In these enlightened days, however, there are many who condemn this kind of teaching as manipulation or indoctrination. The ethics of the alternatives they offer are not always beyond question, but that is not our concern here. What needs to be asked first is whether, and when, manipulation is unethical in this context.

This is a complex question which could take us far beyond our present topic, but one possible suggestion emerges from what has just been said. The basic objection to the manipulator is that he dehumanizes by treating people as objects. We have seen, however, that where a certain kind of reciprocal coupling exists between two individuals, they necessarily cease to be fully specifiable to one another. In that sense it becomes self-contradictory either to treat or conceive of the other as a mere object.[13]

Not all manipulation requires withdrawal from an interpersonal relationship to be complete. What is needed is only sufficient "uncoupling" to provide predictive information in the area of the manipulation. Only that area, and not necessarily the whole individual, is and must be dehumanized for this purpose. It would therefore seem that to define manipulation as *ipso facto* unethical would be to fasten dangerously upon the wrong thing. What is ethically significant, and may be either good or bad, is the kind of one-way relationship necessitated by the manipulation. It is in part the lack of reciprocal openness, the rejection of all answerability by the manipulator, we condemn. That is what makes his attitude and practice unethical.

82

But, it may be said, is not the trouble with the old-fashioned parent or teacher just that his teaching is not open to question, and that he declines to be held accountable by the child he instructs? This may be true enough—at the time. It is equally true of the clinician that he declines to give all the judgments of his patient equal weight with his own. Each rejects questioning for what he regards as good reasons, whether immaturity or mental disorder is in his charge. To this extent fully symmetrical dialogue is in fact impossible for the time being. It would be a naive mistake, however, to conclude from this that these admitted manipulators have rejected all answerability in principle. For at least the best of them nothing would give more pleasure than to be confronted in due course by their charges, mature and of sound mind, and to answer for their manipulative actions in full dialogue. Only, in the nature of the case, this pleasure has to be deferred for a time. They believe that they best betoken their future answerability by exercising the highest skill now, in the most single-minded way, to the purpose that they will eventually be glad to defend is asked.

POTENTIAL ANSWERABILITY

It is the nature and degree of potential answerability in this sense, rather than the degree of manipulation being exercised, that I suggest we should concentrate upon for ethical purposes. If we want to make it more concrete, we can envisage an imaginary *advocatus infantis* confronting the educator here and now on behalf of the child, and demanding all the explanations to which the child as an adult will one day be entitled. It is by the educator's response that the ethics of his manipulations would stand to be judged. Biblically, of course, there is such an *advocatus*, Christ himself, to whom we shall have to answer on the Day of Judgment (Matt. 25:31-46; John 5:22; Rom. 14:10).

Admittedly, in proportion to their capacity for dialogue, the educator has a measure of answerability to the children from their earliest communicative years. More and more forms of manipulation that were initially legitimate become degrading and unethical as the child grows up, because they would negate a level of dialogue for which he has gradually acquired the necessary equipment. It is not that any technique of instruction is legitimate with children provided that we would be prepared to defend its purpose, but the converse: a technique cannot be ethically evaluated solely by asking wether it is manipulative. We must ask whether a manipulation is being used as a substitute for a level of dialogue to which the child's maturity entitles him. If so, it is the refusal of due

83

dialogue rather than the presence of manipulation as such that invites our condemnation.

This last consideration rules out the appeal of the benevolent dictator, who might also claim that in due course he will be prepared to defend his manipulation. A dictator, benevolent or not, is someone who does not at the moment want to be answerable to anyone. His methods are objectionable, not simply because they involve manipulation, but because they avoid dialogue with people who (unlike children or mental patients) are fully capable of and entitled to it now. The category of potential answerability applies only if actual answerability is impossible for good reason.

I have dwelt at length on the example of education because its problems are concrete and familar. But the same principles would seem to apply to human engineering in general. Our question always must be whether, in the sight of God, we are expressing our potential and actual answerability as fully and realistically as possible to our fellow man, and to Christ as his *advocatus*.

PERSONALITY CHANGES

One particular type of manipulation, physical operations to alter personality, raises a special problem. What if a violent criminal, for example, could be changed into a peaceful citizen by otherwise harmless brain surgery? It could be argued that the community would be better off losing an undesirable member and gaining a desirable one, and that the latter (if we have been successful) would in due course be a grateful to us as the Gadarene demoniac was to Jesus (Luke 8:38). The difficulty is that the man to whom we are currently answerable, when considering whether or not to perform the operation, is the violent person we want to change. Can we meaningfully say to him that he will later be glad if we do so? Or is it not he but someone else, the different individual who will emerge from the operation, who will be glad? In the Gadarene case a convenient distinction between the man *per se* and the "unclean spirits" who possess him. It appears Jesus addressed them separately (loc. cit. v. 29). Is there a clue here to a parallel distinction that would help us in such cases?

The problem is closely linked with the philosophical enigma o personal identity: what is it that makes you continuous today with the individual you were yesterday? I know no fully satisfactory answer.[14] However, even to pose the question suggests that in working out the ethics of brain surgery we must recognize two different possible kinds of outcome: 1) those in which the essential brain correlate of individual continuity is preserved; 2) those in which the operation has destroyed brain structures essential to that continuity, in effect terminating the life of one individual and bringing a different one into being.

Assuming an ideal outcome in the first category above, there seem to be no biblical objections in principle against possible chemical treatments that might modify brain functions and dispositions non-destructively. Whatever the differences between the New Testament phenomenon of demon-possession and contemporary psychopathology, the principle of distinguishing between the real man and his affliction, and remaining answerable to the real man while removing the characteristic that afflicts him, seems to apply in both.

It must be emphasized that most brain operations today serve their purpose by destroying brain structures rather than replacing or renovating them. These operations have a serious cost to the patient quite apart from any ethical doubts about the manipulative approach.

We would have to ask ourselves whether the individual now before us ought, in effect, to die. There would be few cases in which we might be convinced that he should. Even in those cases it might be far from obvious that we had any right or obligation to bring a new individual into being in what would remain of the original brain and body after the operation. In other words, assuming it were ever right to bring a particular individual's existence to an end, it might be more ethically defensible to destroy the body as a whole, than to keep it running with a new and alien personality.

One remaining theological question is often raised. Brain surgery can so alter the personality that a lifelong Christian might lose all apparent interest in religion. In such a case, what happens to his soul? The short answer, I believe, is that God has not told us, but we can be sure that He will be more just, realistic and merciful than we would be. If I may offer one very tentative, constructive suggestion, it would be that eternal life is something into which Christians enter here and now as soon as they become united with Christ (John 6:54 and passim). No one who has truly received eternal life can be robbed of it by brain manipulation any more than by death (John 10:29). If I am right in my reading of these and related biblical passages, Christians have no reason to fear the eternal consequences of such an operation. They would have to recognize that it might terminate their experience of God here and now. For practical purposes brain surgery might bring their spiritual life on earth to a premature end.

85

As we have already recognized when talking of education, the shaping of personality and behavior can be undertaken without directly interfering with the physical structure of the brain. The difference is roughly analogous to that between changes to the "hardware" of a computer system by replacing or destroying components and changes to the "program", which leave the hardware intact. In principle it is possible that a program change in a computer could bring about gross physical damage (e.g., by overheating). Analogously, it is possible for purely psychological methods to damage the delicate physico-chemical balance of the human brain-and-body system. But the distinction is a useful one, as there are many ways of manipulating human beings which stop short of gross physical interference. We must briefly ask whether the biblical doctrine of man has anything particular to say about these.

As far as I can see, the possibility of controlling behavior by psychological techniques raises few theological issues that have not already cropped up. The main reason is that the Bible is a good deal more earthy and less high-minded about our bodily endowment than the classical Greek tradition, for example. Man is indeed a "living soul", but he is also "dust."[15] The unity of mind and body in biblical thought is so intimate that I think it makes relatively little difference theologically whether manipulation is carried out at the physical or the psychological level or both. The same stringent criteria must apply.

The idea seems to have found its way into some recent Christian thinking that if a man's mind is in "control," his brain activity and physical behavior must be inexplicable in mechanistic terms. This, I believe, has not biblical support.[16,17] The notion that a complete explanation of human action at a mechanistic level would debunk its spiritual significance, and the agent's responsibility for it, is equally unfounded in logic. In brain research conscious experience and brain activity seem to be harmoniously relatable as the inside and outside aspects of the mysterious unitary process that we know as human agency. Nothing in the Bible conflicts with this. We must be cautious about making artificial separations between our responsibility towards a man's brain-and-body, and towards his soul.

Perhaps the most specific theological issue here is the relation of manipulative techniques to the process of Christian preaching and conversion itself. Much that was said earlier about dialogue and potential answerability applies with double force, since the preacher or evangelist purports to be speaking on behalf of God, introducing his hearers to a personal relationship with God.

What must be recognized is that no matter how genuine and non-manipulative the dialogue between preacher and hearers, there well always in principle be a detached observer-standpoint from which the psychological mechanics involved can be studied in causal terms. To some extent these might be explained by a third party. This has in itself no logical tendency to debunk or devalue the outcome[18]; it does make it possible for would-be manipulators to see what works. They may even develop imitative techniques which are unethical because they do not express genuine dialogue. It also constitutes a temptation for the preacher to lose the posture of dialogue, and subconsciously or otherwise take an unethical observer-attitude to his own performance.

It does not follow, I think, that biblical preachers should stay ignorant of psychology. Rather, they should be alert to all it can say of the dangers of overriding or undermining the mind and will of their hearers instead of illuminating them. If they are fully aware of the presence of God, who sees through them and abhors deceitful dealing of every kind, they have the the best safeguard against misuse of their knowledge. If they are not so aware, no ignorance of psychological technology can save them.

In all this the ethical content of what is being preached or propagated is a separate issue. To foment hatred between man and man, for example, would be abhorrent to the Christian whatever the methods of persuasion used. Our question concerns the legitimacy of psychological means for good ends. The biblical criteria seem to be just as similar and stringent as in any other context.

SOME BIBLICAL CHECK-POINTS

The Christian who seeks to be biblical in his obedience has thus to steer between two opposite evils. One is that of fatalistic complacency: "If God wills that such children should be born, who am I to interfere?" This we saw to be theologically inept. The other is that of naive utopianism, which makes perfectibility by human contrivance to a fallen race is cruelly unrealistic. On the other hand, if Christians are to be the "salt" of the earth they must be in, if not indeed leaders of, efforts to see what good can be done for mankind along these new lines.

How will this "salting" function show itself? There is no
formula, but a short check list of questions may outline the
kind of responsibility that the Bible lays on the Christian
in this area:
(1) Granted the freedom of scientific inquiry, have we properly assessed the relative priorities of the different areas
in which effort could be spent?
(2) In avoiding self-confident atheistic arrogance or wholly
pantheistic optimism,[19] are we sufficiently sensitive to the
claims of compassion?
(3) Do we have a biblically balanced conception of human
needs: the need to give as well as to get; the need to express love and gratitude in worship as well as the need for
security; above all, the need for eternal life, without which
all is ruin in the end?
(4) Have we in mind the Bible's priorities for the subjects
of our intervention?
(5) Have we established a need to act?
(6) Have we done all we can to foresee possible consequences
both for good and for evil, at the three levels of the individual, the family and society?
(7) Is love the ruling spirit of the enterprise?
(8) Above all, are we out to please God, undertaking nothing,
in method or aim which we cannot offer for His blessing as
an act of service to Him.

New knowledge creates new sins, both of commission and of
omission. Temptations to both abound, particularly where
personal reputations are waiting to be made and power waits
to corrupt. Christians cannot pretend that inaction will
absolve them of responsibility, nor can they claim any monopoly of wisdom. Learning what to want is even mor difficult, most of the time, than learing how to achieve what we
want.[20] Christians must be prepared to work hard and humbly
and compassionately at this, alongside all men of good will.
The assurance of the Bible is that the wisdom we need--just
enough, no more--is ours for the asking.

FOOTNOTES

[1]R. Hooykaas, *Religion and the Rise of Modern Science*
(Grand Rapids, Mich.: Eerdmans, 1972).

[2]D.M. MacKay, *Freedom of Action in a Mechanistic Universe*
(Eddingon Lecture; London and New York: Cambridge Univ.

Press, 1967). Reprinted in *Good Reading in Psychology* (M.S. Gazzaniga and E.P. Lovejoy, eds.), (Englewood Cliffs, N.J.: Prentice Hall, 1971.)

[3]R. Hooykaas, *loc. cit.*, Chap. III.

[4]Quoted by R. Hooykaas, *Christian Faith and the Freedom of Science* (London: Tyndale Press, 1957), p. 18.

[5]R. Hooykaas, *ibid.*

[6]C.S. Lewis, *The Abolition of Man* (New York: Macmillan, 1962), Chap. III, esp., p. 48.

[7]R. Hooykaas, *op. cit.* p. 74.

[8]Westminster Shorter Catechism

[9]In what follows I have made extracts from my paper on "Information Technology and the Manipulability of Man," which was published in German in *Zeitschrift fur Evangelishe Ethik*, XII/3, (1968); and in English in *Study Encounter* V/I, (1969), pp. 17-25.

[10]D.M. MacKay, *op. cit.* pp. 121-138.

[11]Deterministic theories of behavior thus offer no excuse from responsibility, since the foregoing argument requires no assumptions to be made as to the existence or non-existence of physical or psychological cues of our actions.

[12]There is, however, an interesting condition in which even an isolated manipulator can be indefinitely embarrassed by his victims, namely if they can equip themselves with an equivalent observing and calculating system, so that they are able to see the outcome of his calculations as fast as he does, and so are in a position to upset the basis of any predictions he might otherwise have made.

[13]D.M. MacKay, *op. cit.*

[14]Four current views can usefully be distinguished: (a) Personal Identity depends on bodily identity (especially of brain). (b) Personal Identity is carried by memory and character. (c) Personal Identity is only evidenced by, but not constituted by, either of these. (d) Personal Identity is a matter of degree. I am indebted to my colleague Professor R.G. Swinburne for the

following references illustrative of each of these views:
(a) B.A. Williams, "Personal Identity and Individuation,"
reprinted in his *Problems of the Self* (New York: Cambridge Univ. Press, 1973), also B.A. Williams, "The Self
and the Future," *Phil. Rev.* 79, (1970), pp. 161-180.
David Wiggins, *Identity and Spation-Temporal Continuity*,
Pt. IV, (Oxford: Blackwells, 1967).
(b) A. Quinton, "The Soul," *J. Philosophy* 59, (1962),
pp. 393-409.
(c) R.G. Swinburne, "Personal Identity," *Proc. Aristot.
Soc.*, (1974), pp. 231-247. R.M. Chisholm, "The Loose
and Popular and the Strict and Philosophical Senses of
Identity," in N.S. Care and R.H. Grim, eds., *Perception
and Personal Identity* (Cleveland: Case Western Reserve
Univ. Press, 1969).
(d) D. Parfit "Personal Identity," *Phil. Rev.*, 80, (1971),
pp. 3-27.

[15]Gensis 2:7.

[16]D.M. MacKay, "Man as a Mechanism," in *Christianity
in a Mechanistic Universe*, ed. D.M. MacKay, (Downers
Grove: Inter-Varsity Press, 1974).

[17]D.M. MacKay, *The Clockwork Image* (Downers Grove:
Inter-Varsity Press, 1974), Chaps. 4, 7, and 8.

[18]D.M. MacKay, *ibid.*

[19]Of the kind popularized recently by Teilhard de Chardin, for example.

[20]On this theme see G. Vickers, *The Art of Judgment*
(London: Chapman & Hall, 1965), and *Freedom in a Rocking Boat* (London: Allen Lane, 1970).

DOMINION OR PAPIER-MACHE?

Robert L. Herrmann
Oral Roberts University

Professor MacKay has given us a superb analysis of the biblical view of man in relationship to his Creator. I was especially impressed by the distinction drawn between man's sinfulness, his refusal to accept God's sovereignty, and his finiteness, inherent human short-sightedness in the face of an
unbelievably complex world.

In the face of such limitations we might be tempted to "contract out" our God-given dominion over the creation, but as
we have been reminded, our opportunity is to be servants of a
Creator who is intimately involved in the agency of His creatures. The world need not be a hostile and alien place, and
nature need not be regarded as a separate "semi-deity" who
"haunts the unwelcome interloper." The biblical God is not
only the divine upholder of the universe, but the day-by-day
encourager of responsible freedom, that man might "enjoy
the peaceful confidence of the servant-son at home in his Father's creation..."

If, as Professor MacKay also points out, obedience to truth
is an essential quality of science, and facts do reign supreme,
then Christians have a special responsibility when looking to

---

ROBERT L. HERRMAN is Professor and Chairman of the Depart--
ment of Biochemistry at Oral Roberts University Schools of Medicine and Dentistry. Prior to that he was Associate Professor
of Biochemistry at Boston University School of Medicine. Dr.
Herrmann has the Ph.D. in Biochemistry from Michigan State
University and was a Damon Runyon Fellow at M.I.T. He is a
member of the American Society of Biological Chemists and a
fellow of the American Association for the Advancement of Science. His research is in biochemical genetics and gerontology.
He has recent chapters in the HANDBOOK OF NEUROCHEMISTRY and
NEUROBIOLOGY AND AGING. He was a member of the Boston City
Hospital Psychosurgery Committee, and is Chairman of the Chris-
tian Medical Society's Ethics Commission.

the future and the application of scientific data to the technology of human engineering.

Reference has already been made to a Boston Conference on
pre-natal diagnosis chaired by Dr. David Allen at which one
of the participants, Vice-chancellor Cooke of Wisconsin made
the statement "Good ethics begins with good data." I take
that to mean that an ethical decision must rely on a full understanding of the ramifications of the choices involved.
Here, where we address ourselves to human engineering, I have
the strong feeling that the data upon which we are to build
such an enterprise is rather limited. It is true that we
have been to the mocn, but the fundamental physics upon which
the space technology has relied was all worked out years before. By comparison, we know relatively little about cells
and organisms in their incredible complexity, and there is
so much more we need to know before human engineering becomes
a suitable subject for ethical decision.

Two examples from my own field of biochemical genetics may
suffice. The statistics of phenylketonuria (PKU), a genetic
disease characterized by the lack of the enzyme phenylalanine
hydroxylase, indicate that approximately 1% of individuals
institutionalized because of mental retardation associated
with PKU have IQ's above 50. Yet these so-called "good PKU's"
have no measurable levels of the defective enzyme. The question which this situation raises is "How good are our biochemical assays and chromosomal analyses in definitely establishing the severity of a genetic disease?" And the more
fundamental question is, "What contribution does the rest of
the genotype, which of course would be different for each of
the PKU patients, make to the expression of the disease?" This
effect of genes upon another gene, which we call a pleiotropic
effect, needs much more study before we can answer these questions. This informational limitation makes it difficult for
the genetic counselor to accurately establish the genetic risk
in a given pregnancy or possible pregnancy. Likewise, the
parent's choice is made less clear and an ethical decision
becomes that much more difficult.

My second example concerns the conference held at Asilomar,
California to consider recombinant DNA technology, the *in vitro*
production of mixed genotypes of bacteria, viruses, and the
cells of higher organisms.[1] The major thrust of the conference
was to set guidelines for experimentation with recombinant DNA
molecules, proposing various methods of containment of possible
biohazards and enumerating which types of experiments were most
hazardous. Reading that report gives one a strong impression
of the complexity of the field and of the wide range of possibilities for both experimentation and biohazard. The field

is fraught with both good and bad possibilities.

Admittedly, scientists tend in general to be overly focused on the good. Indeed, I had the opportunity to recently attend a two-day conference sponsored by the New York Academy of Sciences and Institute for Society, Ethics and the Life Sciences to consider the ethical problems posed by recombinant DNA technology. The impression of several ethicists who had also been at Asilomar was instructive for us. They felt that the ethical problems were scarcely touched upon at Asilomar, and that scientists were largely occupied in sharing the excitement of their work and in anticipating biohazards of the public health type. Senator Edward Kennedy recently made the same judgment. He said the Asilomar meeting was "commendable but inadequate" because scientists alone were involved and because "in fact they were making public policy. And they were making it in private."

But I do not think this possibility, this suspicion, that science might be misused should itself be misused to limit man's exploration of God's creation. Our best function would be to discover how to best provoke a sensitivity on the part of scientists and indeed all possible contributors in the human engineering enterprise to the ethical implications of their work.

My other concern is to emphasize that we need to know much more about pleiotrophy, about interaction of component DNA's in recombinant molecules, and a host of other basic questions. This is at a time when our society is pounding the table demanding results with one hand and withdrawing support for basic science with the other!

From the biblical perspective Professor MacKay has pointed out that science arose in the medieval period, in which learning consisted mainly of knowing the pronouncements of ancient authorities. It arose among a group of largely religious men who somehow regarded the study of ordinary matter worthwhile. Science flowered in the hands of men like Bacon and Newton and Boyle and Huygens, Puritans who believed that the Gospel of Jesus Christ related to every aspect of their being. For them, science was discovery - discovery of the real world which God had made and pronounced as good.

Professor Walter Thorson, in a recent lecture,[2] pointed out that the world in medieval thought was papier-mache, a facade, a thin fabric that you could poke a hole through, for it served only as a brief stopover on the way to eternity, a mere backdrop for the drama of salvation. Yet God had said at each stage of its creation, "It was good" - not just accurate, or uniform, or geometrical, but good! Perhaps the creation mandate to have dominion carries with it much more responsibility in terms of understanding this real world which God takes so seriously. Perhaps, too, this is the reason for that inexplicable "kick" that the scientist experiences with each new discovery, whether large or small.

The heyday of science in our society as a bona fide activity in its own right is over. Now we must be doing something, applying our knowledge rather than seeking new ideas and concepts. Yet it is my expectation that we need much more fundamental knowledge before human engineering can be an ethical activity.

Professor MacKay has emphasized the positive encouragement which the biblical doctrine of creation gives to the scientific approach. He has also indicated how we may see biblical and scientific views of man as complimentary. If science can give us better data for our ethical decision-making, and if science can broaden and sharpen our theological perspective, then, as Professor Charles Coulson once wrote "science is a fit occupation for a sabbath afternoon."[3]

It is tragic that through misunderstanding and often through jealousy, scientists and theologians have gradually gone their separate ways, the scientist frequently forgetting that it was a biblical perspective that forged his beginnings. In light of the present negative attitude on the part of so many institutions in our culture - political, economic, religious - toward science, it would seem to me that an act of Christian charity might be both appropriate and rewarding; namely, that the church once again adopt the furtherance of scientific scholarship through its colleges and universities. My fondest hope is that Christian institutions would see top quality science as an essential goal, as part of educating young people to embrace fully and seriously God's gift of servantsonship.

Should such a hope be realized and science resume its press toward excellence, then ours becomes a double responsibility. If science is to have a strong and free hand on the tiller of its ship of knowledge, then we who are a part of the ship's crew must be responsible to point out the reefs and shoals encountered in these excitingly unfamiliar waters. If, for example, the interaction of different DNA's in recombinant DNA molecules leads, in transformations studies, to the possibility of the injury of human cells, then we who regard human life as infinitely valuable should be advocating additional Asilomar-like conferences in which effects could be examined

in cell and organ culture systems with a view not only
to containment of the possible biohazards, but also with a
view to the examination of the deleterious effects upon the
cell themselves. If gene transfer is ever to be applied
ethically in human engineering, it is essential that we anticipate the harm which may be done to the patient, even if
our test system is crude and over-simplified in patient terms.

But because there is possible harm we should not regress
in our Christian thinking to the medieval universe of papier-
mache. Science has shown us that God has given us a real
world in which to exercise dominion. That dominion has been
enormously productive and we all enjoy its benefits. It is
undeniable that science has been sometimes misinterpreted to
give us a low view of ourselves and sometimes misused to foul
our waters and devastate our enemies. But that is no excuse
for seeking to turn the clock back, even if that were possible. Science, in some form, will persist, because we must
have it. Whether that form will be what the Creator intend-
ed-sensitive, pure, untainted, seeking truth with a passion
and zeal as though thinking God's thoughts after Him - may
well depend upon how we exercise our responsible freedom.

FOOTNOTES

[1]"Asilomar Conference on Recombinant DNA Molecules," *Science*,
V. 188 (June 6, 1975), pp. 191-194.

[2]Science and the Spiritual Dimension, Century 3 Series,
Grace Chapel. July, 1975.

[3]Charles A. Coulson, *Science and Christian Belief* (Toronto:
Oxford University Press, 1955).

TOWARD A NORMATIVE ETHIC

James H. Olthuis
Institute for Christian Studies

Unless the normative meaning of concepts such as life, human, free, responsible, value, love, justice, technology, science is clearly delineated, little progress can be made in stipulating boundaries appropriate to human engineering. Until now consensus on these matters within as well as outside the Christian Church has been, at best, vague and minimal, often more imaginary than real.[1]

A large part of the problem is that technological advances have outdistanced our growth in understanding of norms and values. Mankind finds itself, largely by default, in the service of science and technology. For many science and technology have become their own reasons for existence, autonomous impersonal powers which threaten humanity even as they grant their benefits. The technological steamroller is upon us--for good or evil--and there is nothing we can do.

In this doleful situation Daniel Callahan is undoubtedly right when he describes our present practice as moving "from issue to issue with little regard for general theory, and practically no sense of how one ought even to construct a theoretical solution to the question of responsibility." It is encouraging to hear his claim that Judeo-Christian values "still serve us quite adequately." But, it is less than reassuring to discover that Dr. Callahan did not lay out the content of the biblical norms beyond the usual platitudes. At present biblical norms lead the way in the present confusion and seem to remain at best a hope.

JAMES H. OLTHUIS, B.D., Ph.D., is Senior Member of theology, of the Institute for Christian Studies. His focus is on philosophical theology and theological hermeneutics, and he has a continuing interest in ethics. He has published FACTS, VALUES AND ETHICS: A CONFRONTATION WITH TWENTIETH CENTURY BRITISH MORAL PHILOSOPHY, I PLEDGE YOU MY TROTH: A BIBLICAL VIEW OF MARRAIGE, FAMILY AND FRIENDSHIP, and WORD AND FREEDOM: FOUNDATIONS OF A BIBLICAL LIFE-STYLE.

In my response to Dr. MacKay's paper I will try to draw out the implications of certain remarks in order to help more clearly and sharply delineate biblical norms relevant to human engineering. Four matters will receive attention: The nature of technology itself; the nature of man; God's normative intentions for creation, including technology; and three examples of how I would sort out answers to problems in human engineering.

TECHNOLOGY IS NOT NEUTRAL

My first comment proceeds from the interesting fact that MacKay, although warning against misuse, stressed the positive features of technology for Christians, while Daniel Callahan, although not denying its benefits, has soberly charged us not to be seduced by modern tecnology's tantalizing promise of transcending human finitude. Both MacKay and Callahan have good reasons for their emphases. MacKay desires that Christians acknowledge with more joy and zeal that technology can be a good gift of the Lord--if exercised humbly and charitably. Callahan realizes that although modern technology is often treated as if it were almighty and omniscient, it is in fact a mixed blessing. It is certainly not the source of human salvation. Putting both emphases together we conclude that Christians should promote science and technology as positive functions, while growing numbers of humanists are voicing their disillusionment. At the same time Christians must not idolize science.

The point that I would like to make in this context is: if technology is to glorify God, we must resist the pervasive notion that technology in itself is neutral. Ironically, this notion of neutrality has significantly contributed to its present god-like status. Often we are led to conceive of technology as a more-or-less self-contained tool which must, subsequently, in a separate move be related to values and norms.

Although the confines of this response do not allow a discussion of all the ramifications, I am convinced that approaching the issues in terms of relating (valueless) technology and (extrinsic) values precludes adequate solutions. Technologies are never neutral. As with any method (from the Greek, *hodos*, meaning way), they begin from some view of life or value-system, they go somewhere, and they end up somewhere. Choosing a method, just like deciding to enter a certain door, or selecting a certain bait, severely curtails future possibilities. The results will be the results of that method. Utilizing a net with coarse webbing virtually eliminates the possibility of snaring sardines even as it dramatically heightens the possibility of catching big fish. In the same

98

way the technological method used determines what can and cannot be done. The technology itself is value laden.

When technology is considered neutral, the meaning of technology is found only in the uses to which it is put. Technology itself is considered to be more or less self-evolving, inevitable, irreversible and self-sufficient. Consequently, human responsibility no longer has a real place in technological development *per se*, if reduced to considering how to promote the beneficial and restrict the baneful effects of the technological apparatus. Technology begins to exist for the sake of technique: what can be made is what ought to be made. In this way those involved in developing techniques are able to abdicate any responsibility for use or misuse. Such a view is dangerous in the extreme. Not only is human responsibility and freedom eliminated in the development of technique, not only are there no intrinsic safeguards against arbitrary misuse, but a technology is developed which begins to make its own demands on society in general and man in particular. Technology begins to control man, rather than man controlling technology.

Indeed, fearing that modern technology is out of control, many are raising a hue and cry for external controls, values and and ideals. Technology appears to be an irresistible superhuman power to which we, at best, can only adjust. Jacques Ellul, the noted French Reformed jurist, believes that genuine human freedom is virtually eliminated. Mankind is drugged by technology. Freedom is only apparent, a mirage necessary to hide the truth of the matter. Good human behavior is "that which is called for by the technique, is described by the technique, is made possible by the technique." "One can call everything in our society into question (including God), but not technology."[2]

Today, technology is both accepted (because of its benefits) and feared (because it threatens man's very being). The significance of this paradox can hardly be overstated. If technology is considered autonomous and no super-arbitrary norms for technological development itself are acknowledged, there is no way to develop a viable perspective on technology. In the end we must then either reject technology in the name of human freedom, or accept technology at the cost of human freedom. Either choice is unsatisfactory.

On the one hand, technology cannot simply be rejected. Technical formation and its power are parts of life without which

99

life would close down and stagnate. On the other hand, technology cannot simply be accepted. When technology and its power is believed to be self-sufficient and is its own norm, it jeopardizes human freedom and neglects concerns of justice, and dignity. In the process, it reduces man to just another object to be manipulated.

However, there is a viable alternative which provides a meaningful perspective for technology. In the Christian frame of reference science and technology are not inevitable, irresistible destinies. Instead, they are powers to be used and unfolded in accord with God's intentions for the creation. Without subjection to God's normative Will, technology derails and can only be experienced as the archenemy of human freedom.

When man realizes his calling to guide technological development in obedience to God's Word, technology will never be considered neutral, impersonal and absolute. Rather, technological developments will be judged normative or anti-normative, proper or improper, good or bad.[3] Freedom within technology will be possible and good as man exercises his calling to have dominion over the creation. In this way science and technology will neither be vilified nor deified, but critically championed.

THE ESSENCE OF MAN AS SERVICE OF GOD

The nature of technology itself is crucial. But equally, if not more important in any talk of human engineering is the nature of man. Only when there is clarity on this matter are we in any position to determine what constitutes an improvement in man. Dr. MacKay has led the way here, too. Since "human life is fulfilled in glorifying and enjoying God" in all his relationships, improvement must mean "enhancing men's capacities for... relationship to God and other people."

Discussion at this point often begins to flounder in generalities because it is not further specified. The Scriptures teach not only what people are for (to glorify God and enjoy him for ever), but that what they are for flows from who they are (servants of God). Too often we act as if man's relationship to God is in some degree or other supplementary to his basic humanity as defined in terms of morality, rationality, emotionality and physicality. The Scriptures do not agree with this characterization. For the Scriptures, humanity is at heart faithful or unfaithful service of God. Man is not a creature that has a calling from the Lord among many other obligations. Rather his calling and responsibility is the ineradicable structure of his humanity.[4] At the same time, the one human task has many sides, among them emotional, moral, physical and rational.

It is the biblical view that service and responsibility lie
at the root of what it means to be man. Any technique, regardless of other relatively beneficial features, which destroys
or impairs man's calling to respond from his heart does not
meet the biblical mark. For that is to treat man as an object.
It is manipulation. Such dehumanizing techniques bypass or
shortcircuit the human freedom to respond, choose and program
the "proper" response desired by the manipulator.

Biomedical technologies do treat certain dimensions of human life in objective fashion. This is legitimate and necessary as long as the practitioners do not begin to treat the
whole person as an object, and as long as the measures taken
enhance rather than erode the person's capacity for self-for-
mation. Normally such incursions will only take place following the patient's responsible decision to undergo such treatment. Even in cases where a person, for whatever reasons, has
no sense of responsibility, or where he has lost conscious control of himself in some advanced degree, treatment must be directed to a re-integration of a person's life. This includes
the restoration of the creaturely possibilities of self-for-
mation and choice.

SOME NORMATIVE CRITERIA FOR BIO-MEDICAL TECHNOLOGIES

In suggesting biblical guidelines for biomedical technologies, MacKay emphasized that love must be the "ruling spirit,"
and that, "above all," we undertake nothing, "in method or aim,"
which displeases God. He is, of course, wholly right.

What is not so clear, however, is what love demands specifically, what *in concreto* is pleasing to God? We know that the
command to love God and neighbor is not just one command, but
that it is the root and summary of all men's duties. What is
too often overlooked, and yet is most helpful in giving specificity, is the knowledge that Christ's summary of the Law is
his reaffirmation of God's intentions with the creation from
the very beginning. To love God and neighbor means to execute
human responsibilities to believe, keep troth, promote justice,
think clearly, and to be social, communicative, styleful,
thrifty, sensitive and healthy. These are God's creational
intentions for man in obedience to His will. All these tasks
issue from God's many-sided Word for the creation and man.

At this point, my only real difficulty with Dr. MacKay's
paper surfaces. Perhaps it is only a semantic matter for him,
although it certainly is more than that for many Christians.
MacKay makes a distinction between the *creative* will and the

normative will of God. Crucial to my approach is the understanding that the creative will of God is the normative will.
Love God means concretely taking the "let there be's" of Genesis seriously as normative for today. Certainly Christ regarded the creative will as normative for us in terms of marriage (Matt. 19). There is every reason to believe that the
same holds true for other human tasks. The same Word of God
holds today for the creation as it did in the beginning (cf.
II Peter 3:5-7). Neither did the fall of man destroy the Word,
Will or Order of God for creation, as so many people assume.
Creation fell, but God's intention and order for creation stand
fast and were reaffirmed in Jesus Christ. The unity of the
Love Command and God's Will for creation provides a concrete,
normative framework for developing and appraising technologies.

Crucial to a biblical world-view in these matters is not
only the conviction that human formative power is a gift and
calling from the Lord, nor that there are a pluriformity of
such gifts and tasks, but that all these mandates are to be
exercised and developed in conjunction and simultaneously as
concrete ways of loving God and Neighbor. The Love Command is
not a mystical, idealistic piety, lacking cutting edge and
clashing irreconciliably with the cold facts of life. It has
in fact real teeth which, if applied, help us cut through the
issues at hand.

Negatively, the fact that little or no attention has been
paid to how the powers of technology affect fulfillment of the
total range of man's tasks dislocates and distorts technology
itself. Technology for the sake of technology unbiblically
treats the gift of power as if it exhausts the Love Command by
itself, instead of realizing that power is just one of mani-
fold ways to love God. Positively, concern and attention
that all creational demands are legitimate and require simultaneous and integrated realization open up, lead, and deepen
technological development itself.

The depth-level criteria for development and use of biomedical technology are thus contained in God's multi-dimensional
Word for creation. Specificity of these norms and their applicability is also part of the human calling. In general, techniques deserve development which foster physical and organic
health, enhance emotional sensitivity, promote consistency and
clarity of thought and which, in so doing, open up rather than
close down the possibilities for commitment, communication,
troth, dignity, style, justice and stewardship.

The second grouping of normative considerations especially
require contemporary analysis because of the propensity to

absolutize technological growth. Proper technological
development is dependent in particular on leading by these
"higher" norms. Within the parameters of this norm network
technology works toward enriching and deepening the quality,
depth and integrality of human life. It seeks to make human
existence in all its dimensions less vulnerable. It serves
rather than determines the contours of the human communities,
such as church, school, family, marriage, state, arts and
recreation, business and industry--all of which are God's
gifts to help us fulfill our distinctively human needs. It
contributes to, but does not determine, lifestyle. The
crucial question is whether or not technology promotes or
threatens the quality of human life, rather than whether the
quantity of physical life is increased or preserved. In
decisions involving technology the overriding and weightiest
considerations must be the uniquely human concerns of mercy,
justice, commitment, certainty, faith, hope and love.

Understanding the Love Command as a reaffirmation of
God's Word for creation highlights the special contribution
which Christians can make in the situation at hand. For
the knowledge that the creation was created and continues
to exist by the Word of God comes only through faith (Heb.
11:13). Only in submission to the Scriptures, as the Word
of God written, is it possible to know Jesus Christ in whom
and by whom all things were created (John 1, Heb. 1, Col.
1). Thus, it is only in Christ that the true unity and integration of the diverse creational demands can be known.
Only in Christ is it possible in principle to resist elevating or idolizing one of the demands and considering it
the origin and unity of all the others. In Christ use of
this world in all its rich diversity need not become abuse
(cf. I Cor. 7:31). This is our task: to work out the
meaning of our wholeness in Jesus Christ--our salvation--
in fear and trembling.

ARTIFICIAL INSEMINATION, IN VITRO FERTILIZATION AND CLONING

In conclusion, I would give three brief illustrations of
how attention to the network of norms previously sketched can
guide us in our decision-making. What is important is not
so much whether my judgment in these illustrations is correct,
but that we have norms to guide us in our decisions.

According to the biblical norm for marriage, husband and
wife are joined together as one flesh in a mutual commitment
of troth[5] for "better or worse." In this context the technique of artificial insemination is morally right only when

103

the donor is the husband. The child so conceived is in
every sense fruit of the marital union of the husband and
his wife. However artificial insemination from a donor
other than the husband breaks the mutuality of marriage on
the biotic level, straining and even violating the fellow-
whip of marriage on all other levels.

In similar fashion, *in vitro* fertilization would be
morally proper, provided it can be done safely, when used
by an infertile couple to have a child of their own, if the
transferred embryo is from one of her eggs fertilized *in
vitro* by her husband's sperm. Here again the new technologies serve to deepen marital life through the formation of
a family without hereby violating the mutuality of the marriage. The same cannot be said for situations in which the
implanted embryo comes from an egg not her own but obtained
from a donor, or in which the embryo is transferred to a
"surrogate womb."

In the first two illustrations, the potential, beneficial
effects of techniques (in spite of possible misuses) to
enrich life leads me to consider their development morally
right and proper. In the case of cloning, or asexual reproduction, however, since the procedure itself bypasses the
need for a male and threatens to undermine the family unit
(father-mother-children) given by God for human reproduction
and nuture, I judge that it is morally wrong and that development of the technique should cease.

FOOTNOTES

[1]Typically, the 1975 Report of the National Academy of
Sciences, *Assessing Biomedical Technologies,* notes that the
technologies at issue "force attention to crucial questions
about what human beings are and what they should be" but
hastens to add that the "Committee has not pursued these
questions," although it "believes understanding and
evaluation of these technologies requires their systematic
consideration." (p. 3).

[2]Jacques Ellul, *To Will and to Do,* trans. E. Hopkin
(Philadelphia: Pilgrim Press, 1969), pp. 1, 189, 191.
For a detailed analysis of the various philosophical views
in regard to technology, see E. Schuurman, *Techniek en
Toekomst Asses* (The Netherlands: Van Gorcum, 1972) with
detarled English Summary.

[3]Thus, we need not buy into Ellul's basic pessimism about technology (while sharing his critique) because we do not believe, as Ellul unfortunately does, that technology is irrefutable, inevitable and autonomous. Cf. his *The Technological Society* (New York: Vintage Books, 1964), pp. 6, 142, 428 and *To Will and to Do*, p. 190.

[4]Eccles. 12:13 reads according to the Hebrew, "Fear God, and keep His commandments; for this is the whole of man," not "this is the whole duty of man" as the King James has it.

[5]For a detailed elaboration of the meaning of troth in marriage, family and friendship, see my *I Pledge You My Troth* (New York: Harper & Row, 1975).

PART IV
GENETIC ENGINEERING

Geneticist Robert Sinsheimer explores the concept of
human equality as it relates to genetic modification of
human beings. Taking the view that man is a product of both
nature and culture, Sinsheimer argues that the combined
effect is one of extensive genotypic and phenotypic diversity.
With the advent of new genetic understanding and techniques,
however, the givens of nature no longer have to be accepted.
The question then is, how shall this knowledge be implemented?
Sinsheimer considers some of the risks involved in going
ahead, and some of the difficulties in stopping such research and implementation of genetic knowledge. In the end,
he argues that we will choose to bring about the genetic
equality of human beings. That is, we will establish
genotypic latitudes that will be acceptable; outside of
those latitudes efforts will be made to correct defects in
order to give every one an equal starting point.

Elving Anderson attempts to formulate some questions that
facilitate a different perspective on the problems being
raised. He rephrases Dr. Sinsheimer's initial questions
about equality into two questions. The first centers on how
to bring about full realization of both human equality and
genetic diversity; the second asks if genetic control would
advance equality in the Jeffersonian sense. He then suggests five guidelines in response to the questions: (1)
protect the freedom and responsibility of the individual;
(2) make the genetic control procedures available equally;
(3) explore the meaning of equal opportunity for those with
varying genetic potential; (4) discuss and develop the
criteria for assessing equality; (5) maintain respect for
individual worth despite defects.

Theologian Bernard Ramm suggests that the major dilemma
being raised by advances in biomedicine and genetics is the
lack of a sufficient, precise ethical framework to deal with

the serious nature of these issues because of such issues as irreversibility and uncontrollability. Finally, Ramm makes several observations about genetic control. He suggests that the absence of common consent will make the development of ethical guidelines difficult, but that scientific enterprise must be continually faced with ethical evaluation by non-scientists. He argues that, ultimately, the decision must hinge on the protection of man's God-given dignity.

GENETIC INTERVENTION AND VALUES: ARE ALL MEN CREATED EQUAL?

Robert L. Sinsheimer
University of California, Santa Cruz

Over two hundred years ago Thomas Jefferson wrote the Declaration of Independence which contains the extraordinary lines, "We hold these truths to be self-evident, that all men are created equal and that they are endowed by their creator with certain inalienable rights." This credo has inspired ten generations of Americans and countless others in other lands. Its validity, meaning and implementation have been debated, interpreted and reinterpreted since its appearance. It will undoubtedly be the subject of intense discussion for scores more.

Does this phrase mean that the differences among men, which are surely self-evident as well, are in no wise innate? Do differences arise solely from the influences of environment and culture? Or does it mean that the differences among men, however they arise, are minor perturbations compared to their basic similarities - despite the evident fact that our increasingly specialized society exploits and relies upon those differences?

Does it mean that however different men are, and whatever the origins of those differences, there is a higher power in whose sight they are equal? Should we order our society on

ROBERT L. SINSHEIMER is Chancellor of the University of California at Santa Cruz. Prior to his current position he was Chairman of the Division of Biology at the California Institute of Technology for several years. Dr. Sinsheimer is the author of more than 200 scientific publications. His research interests are the physical and chemical properties of nucleic acids, the replication of nucleic acids, and bacterial viruses. He was chosen "California Scientist of the Year" in 1968 and was awarded the Beijerinck Virology Medal of the Royal Netherlands Academy of Sciences in 1969. In addition he has been chairman of the editorial board for Proceedings of the National Academy of Science and past president of the Biophysical Society. He holds honorary degrees from St. Olaf College and Northwestern University.

that model so as to confer upon each the maximum feasible degree of equality?

Or does it mean that unless we do accept that all men are
created equal - and thereby perhaps define equality - we will
"render Nature herself an accomplice in the guilt of political
inequality", to quote Condorcet? Should one of the functions
of a human society be, within its capacities, to ameliorate
the harshness of Nature?

Some of these questions are factual and can, in principle,
be answered by the methods of science. Some are political or
ethical. These call for other modes of analysis which cannot
ignore the initial facts, even if the aim is to minimize their
consequence.

INHERITANCE AND CULTURE

I would suggest that it is useful, and perhaps self-evident,
to view man as half a creature of nature and half a creature
of culture. There is no equality in nature, except the ultimate grave. There are no rights in nature. Nature uses the
individual; each is expendable. Man exalts the individual and
denies that any are expendable. Rights and equalities exist
only within human societies.

A concept of equality implies a metric measurement, a scale.
Whether or not men are created equal, they most certainly are
not created identical. Each human being is, from the moment
of his conception, the product of his genes. Whatever good
or ill fortune may befall him - *in utero* or after birth - his
genes will help to shape the consequence.

I do not wish to revive the nature-nurture controversy.
Are there aspects of humanity beyond the reach of genes? Yes
and no. Genes furnish the capacity for humanity; they do not
assure its achievement. Genes set the stage for the act of
civilization.

The growth of any individual human being is clearly a highly complex process. It involves the progressive development
and expression of his or her genetic potential in interactions
with our social and cultural milieu. Just as it would be foolish to assume a complete genetic predetermination independent
of the social environment, so it is equally foolish to assume
complete cultural determination, independent of genetic traits.
Rather, genetic factors are significant components in the
determination of both physical and psychological traits. These
include individual sexuality, aggressiveness, temperament, IQ,

110

artistic ability, etc.

THE HUMAN GENE POOL

The gene pool which is the heritage of humanity is re-
markably varied and diverse. This has only recently been
fully appreciated. You may think of your inheritance from
each of your parents as a chain of about three billion nu-
cleotide sites. The chain you inherit from your father will
differ from that you inherit from your mother at about 1,000,
000-3,000,000 of those sites. These differences will be scat-
tered throughout the chains. If we take, as an average, 1,000
sites per gene, up to half the genes you inherit from one parent may differ in some respect from the corresponding genes
from the other parent.

The consequences of this difference may be negligible
or they may be very apparent. Empirically, about 10-15% of
the gene pairs received from the two parents are quite evidently and detectably diverse, heterozygous.

This internal diversity reflects the fact that we each
comprise a minute part of that very diverse human gene pool
which was bequeathed to us by our ancestors and is given by
us to our descendants. The particular set of genes with
which each of us is favored is derived from the two sets of
genes of our parents. Some of us are not so favored. The
laws of chance know aught of compassion, but blindly cast
their dice.

Thus in the normal human being, the three billion nucle-
otide sites are specifically arrayed on 46 chromosomes - 23
pairs. To have more than 46, or less, is in general a mis-
fortune. Yet, at least one in five children conceived will
have 45 or 47; most of these perish in utero. (For special
reasons an extra X or Y or 21st chromosome, or a missing Y
chromosome, is not too deleterious.) Almost 1% of all births
are of this class - all with evident consequences.

For those of us whom the dice have given 46, our genes
will determine whether we are to be fair of skin or dark,
fleet of foot or clumsy, keen of eye or myopic, tall and
lithe or short and squat; whether we are likely to be quick
of mind or retarded; whether we will be robust of health or
frail; whether we will be prone to early heart failure, to
diabetes, to cystic fibrosis, to Huntington's Chorea, to manic
depression, or to schizophrenia, to name but a few of the ills
that flesh is heir to.

111

Genes, of course, preceded society, but the consequences of some forms of genetic variability are accentuated by society. To be lefthanded, which is part of genetic origin, is a handicap in a society which has standardized screw threads and scissors. To be enriched in certain hydrocarbon hydroxylases is probably of little import, unless you happen to smoke, in which case it will at least triple your chance to acquire lung cancer. And, increasingly, our society places a premium upon certain modes of intellectual agility, certain traits of innovation.

It is this enormous diversity of the human gene pool that ensures that each of us is indeed a unique gene combination. Our like will not occur again. All of this, of course, is prior to the cultural diversity in which these varied genetic potentials may flower.

INHERITANCE AND SOCIETY

Are we then created equal? Is the metric of equality established, in part, by the demands and rewards of our social order?

Until now we have been called upon to accept, without question, the outcomes of these fateful games of chance. Our basic genetic endowment and our individual variability have clearly shaped our concepts of the nature of man. Our social structures have been, more or less, well adapted to the array of talents and personalities which have emerged from the existing gene pool and developed through our cultural agencies.

We have, biologically, remained cradled in that web of Nature which bore us. This has undoubtedly provided a most valuable safety net as we have, in our fumbling way, created and tried out varied cultural forms.

But cultures also evolve according to principles we do not always understand. Some people have questioned and expressed grave concern that our evolving technological society is no longer well adapted to human potentials. They fear that it can neither find adequate work for a significant fraction of the population nor provide adequate satisfactions for many. Others claim that we are over-educating many, beyond the capacity of our society to make effective use of their training. Still others suggest that by exploiting and rewarding certain individual qualities we increasingly mock the concept of human equality.

Be that as it may, these disputes frame a real problem.

112

Our culture has brought us to a point at which, if we wish, we soon will need no longer accept our genetic endowment as given. We can expect increasingly to have the means to intervene in the human gene pool in a conscious manner, if we choose to do so.

MODERN GENETICS

I would like briefly to review the recent discoveries that have engendered this prediction and then consider the social and ethical implications of this new vista. Even if my predictions turn out to be exaggerated, or if, for many reasons, man may choose a wholly different course, mere consideration of the possibility of human genetic intervention in the human inheritance affords us the opportunity to examine age old issues of liberty, responsibility and morality in a new perspective. We habituate to and lose sight of the overfamiliar. We have taken certain human conditions as given for so long, that we no longer even see their presence and their role. Our perceptions of possible futures will help us to shape the future.

This prospect arises from the magnificent discoveries in genetics and cellular biology of the last 30 years. The chemical nature of the gene has been elucidated; the chemistry of mutation, and thereby the origin of genetic diversity, is now understood. Much new insight has been gained into the genetic processes and changes underlying the long course of biological evolution.

The complex process of gene expression has been clarified. We can now comprehend much of organismic development and physiological integration as the product of controlled gene expression. The mechanisms of gene control remain, however, as yet poorly understood and they are under intensive study.

There is very much yet to learn. I do not wish to imply that genetics or biology are completed sciences. But accompanying these advances in knowledge and concept have already come advances in gene and cell technology that afford the opportunities for major genetic intervention. These will come first in the simpler organisms.

Let me cite an example. Because of their favorable characteristics, the genetics of certain micro-organisms has been very extensively studied. As a consequence of recent advances we can now transfer genes from other organisms, from plants or animals, from invertebrates or vertebrates, into such micro-organisms and thus propagate these genes within our

113

laboratory cultures. We can also, in principle, insert chemically synthesized genes, such as may never have existed,
into these cells. Such cells can then serve as biological
factories to turn out whatever products we order. It can
be only a matter of time before a wide variety of complex
products of human value are made in quantity in such cells.
These may include insulin, growth hormones, specific antibodies or the clotting factor VIII deficient in hemophiliacs.
Further possibilities of using genetically reprogrammed micro-organisms for industrial or agricultural purposes, such
as a much wider extension of the capacity for nitrogen fix-
ation, are indeed promising.

At this time, the extension of such delicate and precise
technology to higher organisms and to man in particular is
more remote. In part this is so because of the greater complexity of the systems involved. In part it is due to the
present inferiority of our knowledge of human genetics. I
will return to this subject later. However, if one appreciates that it is only in the last eight years that we have
been able unambiguously even to identify each of the human
chromosomes, and if one also appreciates the extraordinary
rapidity of the advances that have been made in other areas
of genetics, one would do well not to over-estimate the time
required to permit significant and detailed manipulation of
human genetics as well.

Other, cruder forms of intervention into the genetics of
higher organisms, including man, are nearer to hand. It is
probably not beyond human ingenuity to develop means to dif-
ferentiate sperm carrying a Y chromosome from those carrying
an X and thereby to provide the basis for the control of progeny gender. The social consequences of a major change in
the human sex ratio, or even of more subtle changes such as
that a large majority of first children might be male, are
difficult to imagine. Certainly they could be profound.

As a general principle, one might propose that whenever
man intervenes to alter a balance provided by nature, then
man must establish his own balance - his own checks and controls, his own constraints. Thus, in seeking to enlarge human freedom we must be careful that we do not in fact diminish it.

CLONING

In another area there has been much talk of the cloning
of animals or man. A clone is a group of genetically identical organisms. As I have mentioned, because of the exceptional

diversity of the human gene pool, clones of humans are unknown - except for identical twins. The similarity of the physical and psychological characteristics of identical twins has often been remarked about. Studies of correlations between identical twins compared to fraternal twins have been of the greatest value in the identification of genetic contributions to many traits and disorders.

The possibility of cloning in higher organisms arises from the circumstance that all the genes of a particular individual are duplicated each time one of his cells divides. Thus the nuclei of each cell in an organism contain its entire genetic complement. Different portions of this complement - different sets of genes - are in use in the different types of cells, but all are believed to be present.

In principle, then one should be able to reconstruct a biological individual from the information available in the nucleus of any of its cells. This has been accomplished in insects and in amphibia.

The most definitive experiments have been performed with frogs and salamanders. The technique has been to extract with a micropipette a nucleus from a cell of an adult salamander, such as a cell of the intestinal lining. That nucleus then is injected into an egg from another salamander from which the egg nucleus has previously been removed. The egg with the nuclear transplant will now begin to divide and go on, in an appreciable percentage of cases, to produce a full grown salamander. This salamander will, of course, be genetically identical to that from which the original nucleus was taken.

If this process is carried out with a dozen nuclei from the same salamander one can produce a dozen genetically identical salamanders, i.e. a clone. The process can be carried on from generation to generation, on whatever scale is desired.

In principle this process should be applicable to any organism. With mammals the egg, with its nuclear transplant, could be reimplanted in the uterus of a hormonally-prepared female to permit gestation.

Such cloning has been attempted with mammals, specifically with mice, but thus far without success. It is possible, of course, that mammals are in some way different and that there is some fundamental biological block. But more likely the difficulty is technical. Mammalian eggs are very small. A frog egg is 3500 times the volume of a mouse egg. The act of nuclear injection is likely too disruptive in mammalian eggs. Various

115

alternative means of achieving the nuclear transplantation are presently under study. No biologist would be surprised to learn that a mouse had been successfully cloned.

Cloning has obvious agricultural potentials, such as the propagation of prize animals. The proposed benign purposes of human cloning are similar to those proposed for other forms of human genetic intervention. Basically there are three rationales: the alleviation of genetic disease, the increase of human self-knowledge, and the "improvement" of the species. I will come later to a fourth purpose, which has been less frequently considered.

The application of cloning to man, once it had become feasible in mammals, would not likely entail any major biological complication. Human cloning would, however, raise a whole complex of social and ethical issues. These range from the propriety of the use of surrogate mothers, to the effect upon the clonées, to the basic social purposes for which such a reproductive procedure might be undertaken.

GENETIC DISEASE

Over the same time interval as these remarkable advances in genetics that I have described, equally remarkable advances have been made in the medical conquest of infectious disease. These have revealed an underlying component of genetic and metabolic disease responsible for widespread human misery, and traceable in whole or in part to diverse genetic factors. I have already cited several of the more prominent examples. Over 2,000 distinct ailments have been linked to genetic influence. The list continues to grow.

It is a simple humanitarian thought to propose that we use our genetic knowledge to attack these inborn errors. Indeed, it can be argued that we are, in the long run, obliged to do so. One consequence of our humanitarian medicine has been to keep alive, and thus to permit to propagate, many individuals whose defective genes would in an earlier era have led to their demise at an early age. This practice, this negation of natural selection, can in the long run, only lead to an increase in the frequency of such defective genes in the human population.

It can be argued that if the defect is treatable, it is no longer a defect and so need no longer concern us. But, as we know, our therapies are often only a partial success. Few would welcome a prospect of lives which require continued and multiple medical treatments.

116

Relief of these inborn errors could assume two forms: prevention of the birth of the affected individuals or gene therapy of such individuals. Gene therapy could ultimately include their germ cells so that the defective genes were not passed on to the next generation. Of these two approaches, prevention is much closer to hand. Somatic gene therapy of an affected individual does not seem too remote, provided certain of the achievements in microbial genetics can soon be applied to higher organisms. Modification of the human gene line does today seem far distant.

Programs to prevent the birth of genetically defective children could take varied forms. Cloning, the asexual propagation of individuals already demonstrated to be relatively free of such traits, is clearly one possibility should it become a major mode of human reproduction. Other possibilities arise from the ability to spot a latent defect in the prospective parents and also from the ability to detect a growing number of defects in the fetus, in utero, by means of the technique of amniocentesis.

Most human genetic defects are genetically recessive - that is, the presence of one normal gene from one parent is ordinarily sufficient, so the presence of one defective gene from the other parent is of small consequence to the heterozygous individual. When two such heterozygous individuals mate there is a one in four probability that they will produce an individual with two defective copies of the same gene, the homozygous recessive.

However, it is possible to detect, biochemically, in many instances, the presence of the carried defective gene in a prospective parent. This leads to the possibility of a sociological solution to the problem of genetic disease for at least those diseases which are relatively common in certain populations. Mass screening could establish which individuals carry the defective genes which produce sickle cell anemia, Tay-Sachs disease or, very likely in the near future, cystic fibrosis. If such individuals would either not marry or not have children by another carrier these diseases (but not the genes) could be eliminated in one generation. Of course, there is a social cost in the (rather small) restriction in choice of mate and in being, as it were, genetically branded.

Amniocentesis permits one to obtain a sample of fetal cells by withdrawing a small volume of amniotic fluid. Fetal cells are normally shed into this fluid. These cells can be cultured and analyzed for chromosomal aberrations or for a variety of biochemical abnormalities.

117

For those disorders which can be detected by amniocentesis technology, there is still another alternative. Even if two carriers of a recessive gene mate, there is but one chance in four that a given child will be genetically defective. The offspring of such couples could each be tested, *in utero*. If a fetus is shown to be genetically defective it could be aborted, if that is considered an acceptable course.

Chromosomal aberrations, or variation in chromosome number, arise as a consequence of accidents in the formation of the germ cells. They are not detectable in the parent. Statistically the likelihood of certain aberrations, such as the extra 21st chromosome which leads to Down's syndrome or so-called Mongolian idiocy, increases rapidly with the age of the mother. If amniocentesis were performed on all pregnant women over the age of 35, two-thirds of all such could be detected and, again, aborted.

GENETIC FORESIGHT

These procedures and proposals carry within them much deeper implications. They are advanced now as means to cope with truly grievous pathological conditions. But as the art and science progress, other questions will inevitably arise. How much genetic foreknowledge do we want? For what conditions is abortion an appropriate act?

Would we, for instance, want to know that a child was very likely to die of heart disease by age 40? Would the child want to know? Would abortion be appropriate in such a case?

Would it be appropriate to abort for purposes of sex selection?

We have already the XYY dilemma. Some individuals, about one male birth in 600, carry an extra Y chromosome. There is valid statistical evidence that the frequency of individuals with an extra Y chromosome is much greater than this among those incarcerated in institutions for the uncontrollably violent. Yet the evidence is also clear that many, if not most, XYY individuals lead socially acceptable lives.

One result of mass chromosomal screening studies of newborns - performed to determine the incidence of various aberrations - has naturally been to identify a number of XYY individuals. What should be done with this knowledge? Should the parents or the child be informed? Can this then take on the character of a self-fulfilling prophecy?

118

Genetic counselors are already confronted with these questions. The answers are not easy.

GENETICS AND HUMAN EXPERIMENTATION

I will return now to the subject of cloning because this serves as an excellent paradigm to explore the widely varied problems exposed by the possibilities of genetic intervention. A second purpose advanced in support of the introduction of cloning is that it will increase our human self-knowledge. Despite much speculation and some analysis, the issues of the relative importance and role of genetic endowment and cultural environment in the determination of a human being remain broadly obscure. The studies of our accidental clones of two, identical twins, have been one of the few effective means of approach to this problem. The use of large clones with carefully selected characteristics, who are then deliberately reared in a variety of environments, might indeed be enlightening. But is it proper to use the lives of human beings in such an experiment as a means to an end? Would it even work? These clones would have to be observed and studied. How would they react to their unusual status? How would the surrounding society react to them?

Even more broadly, is it somehow inhuman to choose a particular role and character for a life? Is it somehow inhuman to design a human being?

We come here to a fearsome dilemma. The methods of science require recourse to experiment. However much we may learn of human genetics from the manipulation of human cells in culture, from retrospective analysis of workings of chance in our unplanned human breeding, and from the most careful study of such favorable cases as identical twins and first cousin marriages, it would seem very unlikely that we can learn enough to predict in detail even the genetic consequences of a program of genetic intervention, to say nothing of the sociological consequences. Thus, the first interveners will be experimenters and the first designed progeny will be guinea pigs. Not all of the experiments will be successful. In plant genetics if some experiments yield a patch of weak crops of low yield, it is simply discarded. What if these were human beings?

The significance of the limitations posed by the ethical quandary surrounding human experimentation should not be underestimated. The traits of greatest interest to those who seek to "improve" mankind are precisely the traits least understood in physiological or genetic terms. These are psychological qualities such as intelligence, foresight, stability,

119

compassion and altruism. The hereditary components of such traits are usually said to be multi-genic, or the consequence of the interactions of numerous genes. However, the evidence for even this vague and plausible statement is considerably less than compelling.

Multi-genic or not, we have no understanding at this time of the physiological and biochemical processes involved whereby genetic elements express themselves in psychological terms. Even if we could tinker with the human genome to re-mold it as we wished, we would be simply ignorant as to which genes to change to what form to influence these qualities.

Nor at this time is it evident how we shall ever obtain such information, except by experiments with their obvious potentials for grievous failure. We may here be confronting a towering road-block to understanding unless future ingenuity finds an adroit solution.

There is another element here. We have long feared, and in the main rejected, human experimentation with human beings, lest once begun we know not where to leave off. This concern has been expressed in such basic principles as the liberty and dignity of the individual, and, in medicine, the doctrine of informed consent. In what name, for what principle, to what end, should one human use another as his research subject?

Yet, of course, in one sense we do an experiment, albeit a poorly planned one, on a human being every time we have a child and draw another hand from that deck which is the human gene pool. The very concept of informed consent is meaningless here. As we know to our sorrow not all of our experiments are successful. The worst failures abort. The others, the sickly and the lame and the retarded, are accepted as part of the human condition, and become our social responsibility.

Part of the answer to Thomas Jefferson is yes, all men are created equal in that their inequality is owed to chance, not to the inadequacies of a human designer. As the innocent victims of chance we do not assign guilt to those unfortunate enough to bear progeny with genetic defects.

But if the knowledge and the means were available to prevent such tragedy, would the parents be blameless if they elected not to use such knowledge? We know enough now that no more children need to be born with Tay-Sachs disease or sickle cell anemia. How should all future parents of such children feel? Is an older morality superseded with newer knowledge?

120

We are repulsed also by the thought of using people as means rather than ends, even if the use is proclaimed to be a means to longer range human betterment. We have had too much experience with the lust for power or even the cruelty of self-deception in human history. Yet, in truth, are means always distinguishable from ends? In a finite world means can predetermine ends and ends can prescribe means. Who really would say that for much of our lives we are not, in fact, means to a social end, even if only the end of individual or group survival? Is a soldier drafted into battle much more than a means? Is the laborer bent in the fields, the worker geared to an assembly line, the typist mechanically punching her keys, much more than a means? The latter, you may say, chose their occupation but what was their range of choice?

It is said that all progress requires risk and pain. But whose is the risk and whose the pain? The question strikes deep into human motive and troubles the conscience. The issue is complex and its discussion may illuminate some of the darker realities of our present social order.

GENETIC IMPROVEMENT

The third purpose presented in the rationale for cloning or for other modes of genetic intervention is the "improvement" of the human race. If it worked, the preservation and multiplication of our finest genotypes would in principle lift the level of the human endeavor. However, let us leave aside for the moment the eminently practical question about whether cloning could achieve the desired result. Let us pass over even the deeper issues as to what really would constitute "improvement" - and who would make that decision. We must first achieve a resolution of purpose before we can judge the means. So, let us ask again, "is it inhuman to design a human being", even to "improve" him?

The initial answer must be clearly, yes. If by "human" we refer to the characteristics of *homo sapiens*, genetic self-design has never been a human prerogative. The values and cultural institutions of mankind have never had to cope with such a possibility. To enter this new dimension would change the nature of humanity, both by the act and the consequence.

Yet, do we celebrate all things human? Hardly. We need hardly be reminded that the human species includes the fool and the bigot, the knave and the cheat, the torturer, the murderer, the tyrant and the slave-master. We need not parade our faults, but few would argue that the human species could not be improved.

121

Is it inhuman to design a human being? To design means to fashion, to shape. We, of course, do mark our progeny now. Our thumbprint is already in our descendants' clay by virtue of our particular genes, our choice of a mate, and the early culture with which we surround them. Our principle means of shaping is through education. Education, of course, informs through the peculiarly human faculties of speech and reason, not through the elemental means of molecules and genes. We tend to find the application of such means to man distasteful. It offends our sense of the uniqueness of humanity. Clearly, one cannot educate a tree or a cow, but one can shape it with molecules and genes. In general we ignore or deplore our animal heritage. But to a biologist even our uniquely human qualities, our capacities for speech, our abilities to learn and to reason, are also an evolutionary gift. We should not scorn, then, to use our reason to re-mold our inheritance.

Is it inhuman to design a human being? The question troubles. It seems at first thought to contradict our cherished principle of voluntarism. It seem to escape the sphere of informed consent. But what does voluntarism mean when in fact there is no choice? No man chooses his genes. The question really is to trust design or to enthrone chance.

Much of human progress can be seen to be the consequence of human efforts to reduce the role of chance in human affairs - the chance of hunger or cold, the chance of attack, the chance of plague. In this endeavor we have not hesitated to intercept the tumbling dice. We have repeatedly mitigated the role of chance and thereby enlarged the domain of human choice and freedom.

But are some matters best left to chance?

For man to seize control of his genetic destiny would indeed enlarge our collective freedom, our collective responsibility, for the two are inevitably coupled.

Some would, with reason, hesitate to impose a new responsibility upon our already troubled social order. We have already a staggering agenda of problems - population, food supply, energy, materials, nuclear terror, environmental pollution. Should we introduce further stress and instability of a most basic nature into the social order which must solve these problems? A sudden major discontinuity in the human gene pool might create a profound mismatch between our social order and our individual capacities. Even a minor perturbation such as a marked change in the sex ratio from its present near-equality could rock our social structures. The impact of a major change in the human life span is hard to predict.

Some would, with reason, doubt that our species has the wisdom to be entrusted with our own evolution.

On the other hand, could it be that genetic change offers the only true solution - the only way out of the dilemma that increasingly arises not from Nature but from man - our flaws and imperfections.

As individuals, our inheritance must forever escape the sphere of informed consent but as mankind, we can bring our inheritance within the sphere of human influence through genetic engineering.

What principles should guide such an endeavor as the genetic improvement of man? If this comes to pass should we adapt our gene pool to some hypothetical ideal society? Or should we adapt our social order to a desirable gene pool? Is our social order infinitely mutable? Or would it be best always to build in some mismatch of human qualities and social structures so as to generate that discontent which is the source of innovation?

In short, how should we deal with such a new dimension of freedom?

I would offer here only a few very general suggestions. First of all, go slowly and as much as possible, reversibly. The human race is very young and there is a long time ahead.

Secondly, preserve human individuality. This is an argument against cloning, except for very special purposes. Our extraordinary diversity may be an accidental consequence of the diversity of the human gene pool, but there can be no doubt that it has shaped our image of the nature of man and underlies our concept of the value of each human being. I, at least, would be reluctant to see that concept eroded.

Thirdly, if and when we know enough, let us augment general qualities rather than specific talents. This will preserve for each individual a wide range of personal option. We do not need to created an insect society. But if we could learn to bequeath to our progeny sounder bodies, more alert minds, freer imaginations, sturdier emotions, perhaps even kindlier, sunnier natures, I would not regard this as bad.

GENETICS AND SOCIETY

In our consideration we should also appreciate that while

123

there are clearly risks to saying "go ahead," there is also a cost in saying "stop." There would be a terrible damage to the human spirit if we should decide that we do not wish to know more, and came to believe that we must forever accept certain ills because we do not trust ourselves with the responsibility to mitigate them. Is there a morality that urges us to refrain, or is there a morality that urges us to act?

We are confronted again with the ancient dilemma of striving versus acceptance, of nerve and daring versus doubt and caution, perhaps of *hubris* versus humility. But the stakes are much higher now. We are no longer innocents cradled in the evolutionary cocoon which bore, sustained and protected us. We are emerging to new potentials for self-expression, to a new level of self-misery.

There is another poignant dilemma upon the horizon. In principle one can beware the Ides of March, but one cannot now escape the wrath of one's defective genes, this innate flaw, this cosmic joke. As our knowledge of human genetics grows, our power to predict bids to outstrip our power to avert. Therein lies the tragedy. In the gift of anticipation man acquired his most powerful adaptive tool and knowledge has multiplied that gift again and again. But anticipation uncoupled to a means of action can lead only to bitter resignation and despair.

The statement "All men are created equal" expresses a very noble human ideal that even the least among us are worthy. That ideal still gleams and still beckons. It was fashioned in an age when man had no choice but to accept as given the outcome of evolution's genetic lottery, to suffer "the slings and arrows" of outrageous chance.

But it would not counter that ideal to attempt to raise the level of the least among us, to repair or prevent those genetic misfortunes that burden so many, to narrow the range of inequalities that nature so heedlessly spawns. Indeed, if we do not, there will be a cancer upon our conscience. As our understanding of the contribution of genetic endowment to individual variability grows, so will this cancer grow. For our society is an incentive society. Its social order is reliant upon differential reward for differential performance. Based upon the credo that all men are created equal, we attribute differential performance to inequality of opportunity and differential motivation. And so we pledge, if we do not provide, equality of opportunity and we make a virtue of motivation.

124

But in our hearts we know that this is a sham. Relative
to the tasks of importance in our society, we are not created
equal. We are not born with an equal chance for health or
wealth, for wisdom or for serenity. That knowledge must gnaw
at our humanity, all the more so if we come to believe we can
lift the unsought curse.

It will come to seem increasingly irrational to tolerate
genetic disability of any sort if we can prevent it. Rather
than distort our definition of equality to fit the callous
diversity of nature we will come to choose our own range of
endowment. Whatever the risk, we will not passively accept
our given lot.

To know how we came to this place and by what route must
then change forever our perception of it. At some fateful
time, in order to enlarge our humanity, we will perceive that
we must accept a responsibility for the genetic endowment of
our children. All the rest will follow, in simple course.

But, we will have need of patience and we will have need of
hope. We grow impatient with the dream of the cultural perfection of man, all the more when our culture itself seems to
be fragmenting under the hammer of incessant change. Having
conquered much of Nature we grow impatient with human nature.
We grow impatient with disease and make "war" on cancer. We
grow impatient with the throws of the genetic lottery and invent genetic engineering. We grow impatient with the human
condition, with our rawness and our finitude. We would be
gods, not knowing the anguish of gods and the dilemmas of
freedom.

We will have need of patience. We should remember that
perfection, like a rainbow, recedes as it is approached.

And we will have need of hope - the hope that imperfect
man can create a finer future - the hope so beautifully expressed by T.S. Eliot:

> We shall not cease from exploration
> And the end of all our exploring
> Will be to arrive where we started
> And know the place for the first time.
>
> T.S. Eliot

THE IMPORTANCE OF THE QUESTIONS IN
ASSESSING GENETIC INTERVENTION

V. Elving Anderson
University of Minnesota

The major objective of my discussion is to assess the relevance that biblical views may have for the issus of genetic
control. Some people have claimed that such views are actually
a major cause of some current problems, such as the population
explosion or the ecological crisis. Similarly, it has been
suggested that a Christian perspective will be ineffectual
(or perhaps even harmful) in preparation for the future.

The nature of our questions are as critical as finding
answers. Some years ago, while leading a discussion on the
Gospel of Luke, I noticed that Jesus often did not answer
questions that were put to him. At first this seemed puz-
zling, but then I realized that he was doing something that
might be more important - he was changing the questions. When
Jesus was asked, "Who is my neighbor?", he told the story of
the Good Samaritan. Then he asked his questioner, "Which of
these men was a neighbor?". The initial question ("Who is my
neighbor?") was intended as a means of self-justification.
The changed question ("Am I a neighbor?") tore through defenses
and arrived at the heart of the matter. Thus, I hope that we
might be able to formulate some questions that will facilitate
a fresh and truly helpful discussion of the difficult problems
we face.

Dr. Sinsheimer's excellent paper raises two central points -

V. ELVING ANDERSON is Professor of Genetics and Assistant
Director of the Dight Institute for Human Genetics at the University of Minnesota. His research interests include the role
of genetic factors in mental retardation, psychotic disorders,
and other human behavioral problems. Dr. Anderson is co-author
of THE PSYCHOSES: FAMILY STUDIES (1973, W.B. Saunders) and co-
editor of DEVELOPMENTAL HUMAN BEHAVIOR GENETICS (1975, D.C.
Heath). He is a regional director of Sigma Xi and treasurer
of the Institute for Advanced Christian Studies.

equality and improvability. The first is much in the public
eye today, as indicated in recent books by Christopher Jencks
et al and H.J. Eysenck.[1,2] The second will be treated by
considering the problems posed by gene transfer.

EQUALITY

Dr. Sinsheimer's presentation dealt very directly with the
meaning of equality in the light of recent advances in genetics.
We must agree that the phrase in the Declaration of Independence, "all men are created equal," has led to confusion and
misunderstanding. It seems likely, however, that Jefferson's
sweeping and unqualified statement assumed familiarity with
the long history of discussions of equality.

Other statements seem to capture the feelings of that time
more adequately. For example, the Virginia Declaration of
Rights (1776) held "... that all men are by nature equally
free and independent, and have certain inherent rights, of
which, when they enter into a state of society, they cannot
by any compact, deprive or divest their posterity." Thirteen
years later, the French Declaration of Rights stated: "Men
are born and remain free and equal in rights." Finally, another Declaration prefixed to the French Constitution of 1793
added: "These rights are equality, liberty, security, property. All men are equal by nature and before the law."

Rossiter summarized the views of that period by saying
that "Revolutionary thinkers were in virtually unanimous accord on this point. Men might be grossly unequal in appearance, talents, intelligence, virtue and fortune, but to this
extent at least they were absolutely equal. No man had any
natural right of dominion over any other; every man was free
in the sight of God and plan of nature."[3] To this Harris
added, "In this sense equality was not absolute and did not
imply identity. All men were equally free, equality and liberty were inseparable, and it was the task of government to
reduce artificial inequalities to a minimum."[4]

Many thinkers have pondered the theme of equality, but it
is of particular interest to investigate the biblical perspective. Sigmund has pointed out that "Christian equality was
not related to a common rationality as in Stoicism but was
based on a common faith in the fatherhood of God, common de-
scent from the first parents and a sharing in their sin, and
a common redemption by Christ."[5] Several verses help to illustrate this point of view:

Ye shall have one law for the stranger and citizen

128

alike; for I am the Lord your God. (Lev. 24:22)
There is no such thing as Jew and Greek, slave
and freeman, male and female; for you are all
one person in Christ Jesus. (Gal. 3:28)
Now there are varieties of gifts, but the same
Spirit; and there are varieties of service,
but the same Lord; and there are varieties of
working, but it is the same God who inspired
them all in every one. (I Cor. 12:4-6)

There is no question but that Christians, along with others, have struggled with this tension between equality and
diversity. A positive addition arises, however, in the realization that these values need not be antithetical in principle. Elton Trueblood has insisted that "One of the best contributions which Christian thought can make to the thought of
the world is the repetition of the reminder that life is complex. It is part of the Christian understanding of reality
that all simplistic answers to basic questions are bound to
be false. Over and over, the answer is both-and rather than
either-or."[6] Frank Stagg has expanded this concept further
by asserting that: "We can find our authentic existence
only in polar situations with their inescapable tension."[7]
An open willingness to face such polarities is neither expe-
diency nor compromise.

Dobzhansky has already pointed out that genetic diversity
and human equality are not alternatives between which we are
forced to choose.[8] Some writers have stressed either of two
extreme positions (which Dobzhansky calls the myths of genetic predestination and the *tabula rasa*). It seems better to
consider human equality as a political and social ideal, and
genetic diversity as a biological reality. One does not have
to deny one to accept the other.

Let me then rephrase Dr. Sinsheimer's question into two
new ones: (1) How can we deal with these two powerful
elements of human equality and genetic diversity in a manner that will permit full recognition of both? A clear response to this question is needed in this time of racial
tension and demands for equal opportunity; (2) If Jefferson were here today I suspect that he would not be surprised
by the growing evidence for genetic differences, for these
were not denied in his use of the word equality. But how
do plans for genetic control help advance equality in the
sense Jefferson intended?

I would suggest the following as an initial attempt at
answering these questions:

(1) Protect the freedom and responsibility of individuals, since "each one must give account of himself before
God." Some proposals have been made for restricting reproduction on the basis of the number or health of prior
children, but such policies would involve a serious limitation of freedom. On the other hand, efforts to protect the
freedoms of special groups such as the mentally retarded
have often failed to recognize that sex education and legally controlled use of sterilization may advance rather than
undermine the freedoms of these individuals.

(2) Make genetic control procedures available equally.
This may require additional public funds and programs of
public eduation as well as some innovations in genetic technology. For example, sophisticated techiques such as those
that might be required for gene transfer will be so expen-
sive that they will be available only to a very few in the
most affluent countries. Schumacher has stressed the value
of "intermediate technology" for third world countries.[9]
In his opinion we need methods and equipment which are cheap
enough to be accessible to virtually everyone, suitable for
small-scale applications, and compatible with man's need for
creativity. What would be an equivalent intermediate technology for genetic intervention?

(3) Explore the meaning of "equal opportunity" for persons with differing genetic potential. The goals of person-
alized education and adapted environment are worthy, even
though difficult to achieve.

(4) Discuss the criteria for assessing equality. Jencks
and his colleagues have been criticized for using income as
a measure of success in assessing the impact of differences
in education in their book INEQUALITY. Jencks responded that
they were merely testing the claim made by others that education is the most effective way of equalizing income. Nevertheless, the point is well taken that income is an ethno-
centric criterion and may have limited usefulness in approaches
to other questions.

(5) Maintain respect for individual worth in spite of
genetic handicap. I support the availability of prenatal
diagnosis, but we must not let this lead to a rejection of
the handicapped or malformed. On the contrary, those who
are more highly endowed or talented have a responsibility
under God to care for others. "I am persuaded that nothing...(not even our genes)...shall be able to separate us
from the love of God, which is in Christ Jesus our Lord."

GENE TRANSFER

Of the various techniques for genetic control mentioned by
Dr. Sinsheimer, gene transfer is especially illustrative.
Clearly the most controversial aspect of this topic would be
the introduction of novel genes, never known to have occurred
in any human. Toffler has echoed some frequently heard speculation on this matter in FUTURE SHOCK:

> We are hurtling toward the time when we will be able
> to breed both super- and sub-races....We shall also
> be able to breed babies with supernormal vision or
> hearing, supernormal ability to detect changes in
> odor, or supernormal muscular or musical skills. We
> will be able to create sexual super-athletes, girls
> with supermammaries (and perhaps more or less than
> the standard two) and countless other varieties of
> the previously monomorphic human human being.[10]

It is not necessary to reject gene transfer outright on the
basis of its not being "human." Many medicines, for example,
are derived from non-human sources. Rather, let us consider
the steps that would be required if gene transfer were to become a reality. First, it would be necessary to understand the
nature of genetic factors that are involved in the development
of specific traits. On this basis alone Toffler's statement is
sheer nonsense. Then gene(s) must be isolated or synthesized
and an appropriate carrier virus selected to convey the gene(s)
into the recipient cells. In order to influence the human
gene pool, the genes must be incorporated into eggs or sperm
cells.

Finally, the genes must be inserted into chromosomes in such
a way that they form a part of a regulatory system for turning
genes on and off at appropriate times in development. This
point was emphasized by Markert and his colleagues in a recent
paper on the enzyme lactate dehydrogenase: "What distinguishes
one vertebrate from another are primarily changes in the timing
and expression and the relative amounts of the same gene products, not the minor differences evident in the structure of
these gene products." At the present time the nature of such
regulatory systems in mammalian cells is poorly understood.

As an aside, this high order of complexity permits one to
reflect on the value of the human body. It has been claimed
that an individual is "worth" about $2 (or perhaps $3 with in-
flation). But this is merely the value after a body has been
reduced to ashes. "All the king's horses and all the king's
men" cannot at present replicate the various molecules in their

three-dimensional arrangement. The human body is truly priceless!

There remains the difficult task of deciding what would be a useful novel gene. The winnowing process of natural selection already has picked out those combinations of genes that work together successfully. The safest procedure would be merely to add an extra dose of a gene already present in the human gene pool through genetic duplication.

Furthermore, any newly introduced gene would almost certainly cause a number of deaths before changes in other genes provided a suitable genetic environment. At the very least I would argue that such gene transfer be explored thoroughly in domestic animals before its use in man is considered seriously.

I now return to Dr. Sinsheimer's question: "Could it even be that genetic change offers the only true solution – the only way out of the dilemma that increasingly arises not from nature but from man – from our flaws and imperfections?" Solution for what? Why the only solution? At what cost as compared with other options? We must agree on the question before we propose a solution, for if our diagnosis of the human predicament is wrong, the treatment will be ineffective or even deleterious.

Peter claimed that "His divine power has given us all things that are necessary for life and godliness." (I Peter 1:3). In some sense at least, then, a means of meeting important human needs is already available. This by no means implies that genetic control is not important, but it does suggest that we ought not to expect genetic technology to solve all human problems.

As a final word, I would underscore Dr. Sinsheimer's call for hope and patience in facing the future. It is said that Abraham, when he was called, went out in faith, not knowing where he was to go. With all of our science and technology, we do not really know where we are going, but we too can go forward in the faith of Abraham.

FOOTNOTES

[1]C. Jencks; M. Smith; H. Acland; M. J. Bane; D. Cohen; H. Gintis; B. Heyns; and S. Michelson. *Inequality. A Reassessment*

of the Effect of Family and Schooling in America (New York: Basic Books Inc., 1972).

[2]H.J. Eysenck, *The Inequality of Man* (London: Temple Smith, 1973).

[3]C. Rossiter, *Seedtime of the Republic. The Origin of the American Tradition of Political Liberty* (New York: Harcourt, 1953), p. 292.

[4]J. Harris, *The Quest for Equality* (Baton Rouge: Louisiana State University Press, 1960), p. 15.

[5]P.E. Sigmund, "Hiearchy, Equality and Consent in Medieval Christian Thought." In *Equality*, ed. J.R. Pennock and J.W. Chapman (New York: Atherton Press, 1967), p. 139.

[6]D.E. Trueblood, "The Self and the Community." *Quest for Reality: Christianity and the Counter Culture*, ed. C.F.H. Henry (Downers Grove, Ill.: Inter-Varsity Press, 1973), pp. 39-40.

[7]F. Stagg, *Polarities of Man's Existence in Biblical Perspective* (Philadelphia: Westminster, 1973), p. 42.

[8]T. Dobzhansky, *Genetic Diversity and Human Equality* (New York: Basic Books, 1973).

[9]E.F. Schumacher, *Small is Beautiful* (New York: Harper & Row, 1973).

[10]A. Toffler, *Future Shock* (New York: Random House, 1970), p. 179.

[11]C.L. Markert, J.B. Shaklee and B.S. Whitt, "Evolution of a Gene," *Science* (1975), 189:102-114.

THE COMMON CALL FOR HUMAN DIGNITY

Bernard Ramm
Simpson College

Sinsheimer pinpoints in genetics the kinds of problems
modern science and its technology have been creating for
ethics for more than a century. Bertrand Russell expressed
the dilemma years ago in his work, A FREE MAN'S WORSHIP.
Omnipotent matter rolls on and crushes man but man is one
up on omnipotent matter: he knows he is being crushed!

Recently Ernest Becker has put it more painfully and more
pessimistically![1] Man is an organism among other organisms
on the crust of the earth. He is eating other organisms at
an enormous rate of consumption until he is eaten or dies.
Human beings differ from these other organisms in that they
have figured out that they are such an organism. The picture
is, however, so unbearable that people have magnificently
rationalized it to bear it. Human rituals, mythologies,
religions, metaphysics and perhaps even universities, are
man's ways of making the fate he grasps with his reason
tolerable to his feelings.

Sinsheimer is not that pessimistic, but Russell and Becker
are looking at the same dilemma from different perspectives.
Man is an unusual organism among organisms. He cannot escape
ethical considerations stemming either from his own actions
or from those of others. Nor can he resist making value
judgments. It is very common to read many sharp value judgments in communist publications. Communism, of course, is
an ideology based on the primacy of science in all matters
of fact. Somehow values have become great matters of fact.

BERNARD RAMM is Professor of Christianity at Simpson Col-
lege, Modesto, California. Dr. Ramm is Theologian in resi-
dence at the First Baptist Church of Modesto and Visiting
Professor at the Eastern Baptist Seminary. He was formerly
Director of Graduate Studies in Religion, Baylor University
and Professor of Theology, American Baptist Seminary of the
West. He has authored many books including THE GOD WHO MAKES
A DIFFERENCE, THE EVANGELICAL HERITAGE and THE CHRISTIAN
VIEW OF SCIENCE AND SCRIPTURE.

Man also makes claim to knowledge. This organism with ethics, values and knowledge is faced at the same time by the great imponderables which modern science, modern technology and modern genetics set before it. The burden is close to unbearable. Scientific advances proceed at more than a geometric ratio. Man's growth in ethical perception is less than a snail's pace.

Even one so cosmically skeptical as Becker reasons that if we repress the conflict (which I gather is the usual course of action in centers of research) we do not solve it. Rather, we make the day of reckoning even more dreadful.

We must admire the courage of Sinsheimer in tackling the issue of ethics and genetics. I cannot fault his guidelines, namely: (1) to proceed slowly in genetic experimentation; (2) to make human individuality a prime value; and (3) to urge advances in general human welfare rather than focus on specialized functions such as a super-breed of soldiers or athletes.

The truth of the matter is that the problems are so complex none of us are very smart. The expert scientist feels a cloud drift into his mind when he moves from science to the ethical principles which should guide his science. A rigid, biblical ethicist paws in the darkness along with the rest of us. But if the dilemma does not go away we have to tackle it by one means or another. I will express my views in three stages: the dilemma, the seriousness of genetics, and some Christian observations.

THE DILEMMA

Russell, Becker and Sinsheimer have stated the dilemma in their respective ways. In summary, it is the severe tension between the rapid and sophisticated increase of knowledge in genetics and the slow, ponderous manner progress is made in ethics. We have no way of "doing ethics" that compares with the controls, experimentations, and sophistication with which one "does genetics." Becker's picture of the dilemma is humiliating but graphic.

THE SERIOUSNESS OF GENETICS

As dramatic as open heart surgery, open cranial surgery, and organ transplants are, they are not within a billion miles of the seriousness of experimentation in genetics. In the former doctors are dealing with one life. Granted, that one life is serious and, to the patient, his operation is the most serious matter in his life. In contrast, genetics

deals with the building blocks (DNA) of life itself. Geneticists are not in the process of saving one life but
are working with the blueprints of life itself.

The most serious matters in genetic experimentation are
irreversibility and uncontrollability. In genetics experimentation can go down a path on which it may not be able
to reverse. Or, new materials may be created for which
there is no defense. The latter has been a matter of great
controversy recently among geneticists. We are in no position to decide the issue. But we can apprehend why some
geneticists are so violently upset just with the specture
of a bacteria of a lethal nature for which we have no antibodies.

In summary, the ethical issues in genetic experimentation are of maximum seriousness both to Christian and non-
Christian alike.

SOME CHRISTIAN OBSERVATIONS

First of all, as a Christian interested in medical ethics in genetics I wish to make some general comments coming
out of my stance.

We are called a permissive society and a society of alternative life styles. To my surprise I saw a large section of a bookstore with the caption "Alternative Life
Styles." The pull of a pluralistic society, a society of
alternate life styles is towards the "least" common denominator.

However, I do not think we shall find our way in science and ethics by the route of common consent or common
opinion or ethical plebescite. I think that the ethicist,
Paul Ramsey, is right in his stance that ethical decisions
must be the product of rigorous logical thinking, fudged
logical processes or principles settled by popular appeal.

There is no neutral, purely technological science, and
therefore no neutral, purely technological genetics. It
sometimes distresses scientists to have ethicists, religionists, theologians and other self-appointed monitors of
public morality prying into their matters. No doubt "snoops,"
"dogmatists," and the ill-bred have done much to discredit
serious ethical reflection in science.

However, the fact that man is an ethical creature and is
capable of right and wrong, that he is a creature of values

137

and therefore is capable of good and evil, and that he is a thinking creature capable of truth and error cannot be avoided in science. "What goes on there?" is always, therefore, a proper question. We dare not stop asking it of the politician, the revolutionary, the corporation executive, the educator, or the scientist. As long as men are capable of evil, sin, and error the question is always legitimate. Geneticists are no more exempt from the interrogation than any other class of citizens.

From the standpoint of positive Christian affirmation I would say the following:

(a) All of humanity is in the dilemma of ethics and genetic experimentation. From the Christian perspective humanity is God's humanity even though we may divide it up into the City of God and the City of Man. Christian and non-Christian must work shoulder to shoulder on issues with such massive implications. Genetic experimentation is experimentation with mankind. It is therefore an issue of the total family of God.

(b) In agreement with Sinsheimer, I think a Christian would also say "go slowly, go cautiously." Whoever would have dreamed that from Becquerel's discovery of mottled X-ray plates and Einstein's E = mc^2 would come the terrors of atomic warfare? One of the most difficult matters in medical ethics is the inability to predict what happens from a certain procedure ten or twenty years later. For example, it was recently discovered that the radiation of the throats of youngsters led to cancer of the thyroid in adult life. If this is a headache in medical ethics, we cannot even dream of the potential monstrous possibilities in genetic experimentation.

(c) As a Christian I would be opposed to any obvious use of genetics or genetic information or genetic experimentation the purpose of which is evil, sinful or selfish.

(d) As a Christian I would further say that in all things a Christian is to be guided by love, redemption, healing, wholeness and reconciliation. Wherever genetics can alleviate human suffering Christians should have a positive attitude. Furthermore, wherever our knowledge of genetics can alleviate suffering or cure infirmity or prevent monstrosity or eliminate great evil the Christian should look upon such with favor. As much as Christians may differ in specific cases I do not see how they can differ in policy.

138

(e). I would not appeal to all men being created equal as the point of ethical departure. The concept appears to me to be a mixed chunk of deism, rationalism and democratic idealism of the eighteenth century. I would take my stance on the dignity of man, which stems from man made in the image of God.

In most ways we are not equals. In contrast, because we are in the image of God we all bear the dignity the term implies. We could even speak of the "redemptive dignity" of man. If God so loved man that Christ died for man there is a revelation of the dignity of man in this we could not know from other sources.

Helmut Thielicke uses the term "alien dignity" in much the same way. By "alien" he means a dignity given man. He has been made wary by the events of the twentieth century to locate the dignity in man himself. He thinks man as man will only be adequately protected in modern medical ethics and genetic experimentation if he has a dignity from God. Humanitarianism is not enough to carry the burden of the issues.

It is not easy to go from the concept of the dignity of man to a specific decision in genetic experimentation. But Christian and non-Christian need some compass. For the Christian I see it as the dignity of man bestowed upon him in creation and redemption. It appears to me the only point of leverage in all of ethics. Man is to be treated as man in his God-given dignity: man in the image of God, sullied or redeemed.

FOOTNOTES

[1]Ernest Becker, *Escape From Evil* (New York: Free Press, 1975).

PART V
ELECTRO-CHEMICAL INTERVENTION

Elliot Valenstein opens this section by reviewing a number of the assertions and claims that have been made about physical modification of the brain. He then proceeds with an extremely helpful and thorough analysis of what the actual facts are, in an attempt to demythologize the fears and claims being sounded. Valenstein points out, for example, that Delgado's classic demonstration of ESB being used to control aggression can actually be explained by reference to motor control. He points out that the evidence indicates aggression is not controlled exclusively by any specific brain site. Furthermore, he cites research showing that brain stimulation produces responses that are different from natural motivational states. Environment also affects the kinds of behaviors expressed in response to electrical stimulation. With regard to psychosurgery, Dr. Valenstein sternly warns against tunnel vision which blinds scientists to the real scientific information in their haste to find solutions to problems. He points out a number of examples in which this has been done. The result is an overly anxious and dramatized response from lay persons. Throughout his paper, Valenstien points out the complexity of the relationship between brain functioning and environmental factors. He argues against an overly simplistic approach.

Finally, Dr. Valenstein argues against scientific moratoria because they impede our ability to acquire the very information needed to make properly reasoned ethical decisions. At the same time he urges the development of rigorous review procedures that insure more objectivity than the present peer review system allows. He also urges that ethical criticisms need to be critically examined as much as the scientific activity under scrutiny. He shows how pure motives and ignorant decision making can lead to the damage of both individual scientists and the scientific enterprise in unjustified ways.

Duke University psychiatrist William Wilson argues that, in spite of potential abuses, the healing and helping of humanity through brain intervention must be actively be pursued. He suggests that the process of investigation in these areas is similar to the Abrahamic step of faith necessary for dominion to be exercised over all the earth. Wilson argues that we should denounce irresponsibility but responsibly use truth. He then shows the value of using certain human engineering procedures, both for the individuals involved and for others suffering from similar problems. Finally, Dr. Wilson pleads for more care in the construction of laws, intended to safeguard individual freedom, which needlessly prevent the use of beneficial techniques in relieving problems.

The second response to Dr. Valenstein's presentation is by theologian Paul Feinberg. Dr. Feinberg focuses upon several fundamental questions raised by chemotherapy and psycho-surgery. First, he argues that the complexity and absence of uniform results that Valenstein points out may very well be due to the nature of persons as presented biblically, rather than due to imprecise methods and instruments. The materialistic committment of many scientist may blind them in their interpretations. Second, Feinberg suggests that *in principle* justifications should be questioned. They may be simply expressions of optimism based on presuppositions that do not actually fit the facts. He then deals with two further issues about the legitimacy of control. He points out several serious ethical objections to the cost/benefit decision-making models often employed.

Finally, Dr. Feinberg argues that the biblical account of creation and God's provision of moral choice--the opportunity to sin--argue against the development of human engineering techniques which destroy choice. This is so, even if the techniques are being developed to prevent ethically bad behavior. Such techniques, he says, deprive human beings of a God-given capacity, and are thus morally wrong.

BRAIN CONTROL: SCIENTIFIC, ETHICAL
AND POLITICAL CONSIDERATIONS

Elliot S. Valenstein
University of Michigan

It is obvious that the new techniques for physically
modifying the brain are fascinating.[1] Newspapers, maga-
zines, novels, movies, radio and television programs por-
tray over and over again the potential of electrical and
chemical brain stimulation to control behavior, emotions,
and even thoughts. Over a million people have read or seen
the movie version of Michael Crichton's novel, THE TERMINAL
MAN.[2] Many of these are now convinced that computer-pro-
grammed brain stimulation can (or will soon be able to)
control human behavior with great precision. A story in a
widely circulated magazine even described a future government, an electroligarchy, in which the social classes would
be distinguished by the number of brain electrodes implanted.[3]
According to the article members of the lowest class, the
"neutrons," would have 500 electrodes inserted, no free will,
and in fact "...be completely robotized. They could dig
ditches all day and love every minute of it." It is almost
impossible to escape from the omnipresent accounts of brain
control techniques.

ELLIOT S. VALENSTEIN is Professor of Psychology and Neuroscience at the University of Michigan. He is currently
President of the Division of Comparative and Physiological
Psychology of the APA and on leave as a Fellow at the Center
for Advanced Study in the Behavioral Sciences. Dr. Valenstein
is the author of more than 100 scientific articles and two
books, BRAIN CONTROL and BRAIN STIMULATION AND MOTIVATION.
His interests range widely in the brain-behavior field, with
research emphasizing the physiological basis of emotions,
motivation and learning. Dr. Valenstein is a member of many
professional societies, including the International Brain
Research Organization and the Society for Neuroscience. He
has served on numerous scientific advisory panels for NIH,
NIMH, and NSF, on editorial boards of many journals, and has
recently surveyed the practice of psychosurgery for the
National Commission for the Protection of Human Subjects in
Biomedical Research.

143

Of course many of the stories are only meant for our
amusement, but that does not stop even these, in time,
from having a profound influence on our processes. It is
becoming increasingly common, for instance, for psychiatric
patients to claim that their thoughts and emotions are being controlled by devices that were implanted in their
brains while they slept. Actually, these paranoid thoughts
are not very different from the ideas expressed by many
normal people. One prominent professor of psychology, for
example, has described the power of electrical stimulation
of the brain in the following manner:

The number of activities connected to specific
places and processes in the brain...is simply
huge. Animals and men can be oriented toward
each other with emotions ranging from stark terror
or morbidity to passionate affection and sexual
desire...Eating, drinking, sleeping, moving of
bowels or limbs or organs of sensation grace-
fully or in spastic comedy, can all be managed
on electrical demand by puppeteers whose flaw-
less strings are pulled from miles away by the
unseen call of radio and whose puppets, made of
flesh and blood, look like 'electronic toys' so
little self-direction do they seem to have.[4]

It would not be difficult to provide many similar examples
taken from widely read books and even college texts. Such
exaggerated and distorted descriptions of the power of
electrical stimulation and other physical techniques used
in brain research have produced not only fears about their
possible misuse, but they have also led to proposals for
applying this presumed power to the solution of social problems. The belief that brain interventions might achieve
desirable social goals was expressed by the social psychologist Kenneth Clark in his Presidential Address at the
American Psychological Association meeting in 1971. Clark
startled the large number of people in the audience by his
statement that:

...we might be on the threshold of that type of
scientific biochemical intervention which would
stabilize and make dominant the moral and ethical
propensities of man and subordinate, if not eliminate, his negative and primitive behavioral tendencies.[5]

In the same address, Clark also suggested that political
leaders should be required to: "...accept and use the

144

earliest perfected form of psychotechnological, biochemical intervention which would assure their positive use of power and reduce or block the possibility of using power destructively." The fact that such a suggestion came from a member of a group that has traditionally favored environmental rather than biological explanations is an indication of extensive influence these distorted descriptions of the power of brain control techniques have had.

In addition to debate over the future social application of techniques to physically modify the brain, there is also controversy currently raging over what has been called the "resurgence of psychosurgery." The use of brain surgery to treat the mentally ill has raised a multitude of ethical, social, and political questions about how best to protect patients and our civil liberties without blocking all experimentation and progress in science and medicine.

SEPARATION OF FACT FROM FANTASY

In order to discuss the problems raised by the new brain techniques realistically, it is essential at the outset to separate fact from fantasy. Distorted conceptions of the power of these techniques can lead to legislative remedies directed at the wrong problems. If in the process unworkable bureaucratic controls are established the result may be very harmful. A closer examination of the results of typical experiments which have been distorted may be the best antidote for the more common misconceptions over the power of brain control techniques.

Several years ago, Jose Delgado described the results of a demonstration of the use of brain stimulation on wild bulls. The situation was so unusual and intrinsically dramatic that the demonstration has been described repeatedly in the popular media and in psychology textbooks. The following account is not atypical of these reports:

Dr. Delgado implanted a radio-controlled device deep within the brain of a brave bull, a variety bred to respond with a raging charge when it sees any human being. But when Dr. Delgado pressed a button on a transmitter, sending a signal to a battery-powered receiver attached to the bull's horns, an impulse went into the bull's brain and the animal would cease his charge. After several stimulations, the bull's naturally aggressive behavior disappeared. It was as placid as Ferdinand. (N.Y. Times, September 12, 1971).

145

Although the interpretation implied in the above quotation is commonly accepted, there is actually little evidence to support the belief that the stimulation specifically inhibited the bull's aggressive tendencies. It is immediately apparent to those who see the film of this demonstration that the stimulation forces the bull to turn in circles. If the stimulation is administered every time the bull charges, the bull seems to become confused and stops charging at least for a short period. This demonstration did not suprise anyone knowledgeable about the nervous system because the stimulating electrode had been placed in a brain structure (the caudate nucleus) known to play an important role in regulating bodily movements. Patients receiving stimulation in this same brain structure typically display various types of stereotyped motor responses.[6] Also, movement disorders such as the spasticity and the tremors seen in Parkinson's disease have frequently been associated with caudate nucleus pathology. No convincing evidence has been presented to support the belief that the bull was pacified in any significant way by the stimulation. Judging by appearances, the bull seemed confused and perhaps also frustrated. In any case, it is misleading to convey the impression that the brain stimulation was modifying aggression exclusively.

Aggression is certainly not controlled exclusively by the caudate nucleus or even the hypothalamus, a brain area where electrical stimulation can provoke normally peaceful cats to kill rats. Experimental destruction of either of these brain areas does not eliminate natural aggressivity unless the region destroyed is so extensive the animal is generally incapacitated and capable of little behavior at all. In any given instance, the interpretation of the behavior evoked by brain stimulation can be quite complex, but it is highly doubtful that the demonstration indicates a way to eliminate aggression while leaving other behaviors undisturbed. A great amount of experimental evidence makes clear that many brain areas can influence the display of aggression. Each of these areas can also influence other behaviors as well.

Another example illustrates a similar point from a somewhat different perspective. The power of brain stimulation to control behavior is often exaggerated because of the belief that it is possible to arouse distinct motivational states by stimulating specific areas of the brain. It is often claimed that the fact that brain stimulation causes animals to eat, drink, engage in more rapid or extensive sexual behavior, carry young, and so forth, demonstrates

146

the possibility of turning such motivational states on and off with a great amount of reliability and selectivity. It is important to examine this conclusion more critically.

In my laboratory we have shown in many different ways that brain stimulation does not duplicate natural motivational states.[7,8,9] It now seems unlikely that the animal that eats, for example, in response to brain stimulation does so because the stimulation has evoked a natural state of hunger. My collaborators and I have shown that even though a rat may eat food when its brain is stimulated, it usually eats only one type of food and not another. The stimulated animal may not even eat the same food after it has been changed in shape or texture, such as when it is offered as mash instead of as pellets. Animals that are known to be hungry because they have been deprived of food do not hesitate to eat the mash. Particularly difficult for the usual interpretation of many of the demonstrations is the fact that the behavior evoked by identical brain stimulation can change over time. Initially, a rat may only eat when stimulated. Later, however, it may drink when the same brain area is stimulated, or it may engage in some other behavior. Moreover, changes in the animal's environment often result in very different behaviors being evoked by the brain stimulation. These facts make it impossible to maintain that brain stimulation can reliably produce the same behavior or motivational state all the time.

If the response to brain stimulation is variable in inbred rats, it is certainly much more variable in monkeys, apes, and man. In humans, stimulation may evoke general emotional states that are somewhat predictable. That is, the activation of certain brain areas tends to produce unpleasant feelings while the activation of other areas tends to produce pleasant emotional states. In response to brain stimulation, patients may report feeling tension, fear, or anger, or they may describe feeling relaxed. On occasion they even become sexually aroused. However, the same patient may have very different responses to identical stimulation when it is administered at different times. It simply is not true that stimulation of a given brain structure evokes the identical emotional state repeatedly.

If stimulation of a particular brain area does not always evoke the same emotional state, it clearly will not provoke the same behavior consistently. It may be convenient and dramatic to speak about the correspondence of brain areas to complex behaviors, but as usually stated this implies

147

a much more direct relationship than actually exists. The
behavior that is expressed in response to even the same
emotional state will vary with the personality of the subject and the situational determinants that are operating
at the time. One person may lash out at the world when
angry while another person may simmer quietly or redirect
the anger into constructive behavior. Certainly there is
little direct correspondence between brain areas and the
desirability of any behavior. We may be concerned about
aggression, but it should be obvious that a concept such
as aggression is a man-made abstraction that does not exist
as a separate entity in the brain. Much of what comprises
undesirable aggressive behavior is undoubtedly very equally
essential to highly desirable behavior. The belief that
we can modify the normal brain and selectively decrease
the expression of asocial aggressive behavior without influencing desirable capacities in any way is not realistic.
This argument is not contradicted by the fact that there
are well documented cases of people becoming assaultive
after brain damage and whose general behavior was more
socially acceptable following corrective neurosurgery.
Even in such instances behavior other than the tendency to
be assaultive is affected, but the overall change may be
judged to be beneficial.

Not everyone will be reassured by the arguments that the
control over behavior by physical interventions into the
brain is much less predictable, reliable, or natural than
usually described. There are those who would argue that,
in view of the progress made so far, it is likely that our
ability to control behavior by physically altering the
brain will be much more precise in the future. While anything is possible, the present data certainly do not suggest
that precise and predictable control over behavior will be
realized in the foreseeable future. Greater accuracy in
placing electrodes in the brain, for example, is not likely
to reduce the unpredictability of changes in complex behavior. The behavior influenced by stimulating any brain
structure (even a single nerve cell) will always depend
upon on-going activity in other brain areas. Undoubtedly,
more efficient means will be developed to alter mood and
to modify mental alertness and memory. This is very different, however, than controlling the specific content of
thought processes or the changes in behavior that will follow modification of mood or cognitive abilities.

The belief that there are brain techniques that can
control behavior in highly predictable ways has generated
reaction in two directions. The fear that this presumed

power may be used as a political weapon has caused some
people to recommend what appears to be very hastily conceived and potentially dangerous legislative action.[10]
Others have apparently been persuaded to abandon the search
for social remedies to problems because of the belief that
there are (or will soon be) biological solutions available.
It would be difficult to fabricate a better example of the
distortions that can result from an exaggerated and unrealistic belief in the capacity of biological techniques
to control behavior than that contained in an article by
Ingraham and Smith.[11] These two criminologists suggested
that techniques are available for maintaining surveillance
on paroled prisoners, and for controlling their behavior.
They described implanted devices for monitoring the loca-
tion and physiological state of a parolee and the possibility of using remotely operated brain stimulation to control
undesirable behavior. The following scenario is described:

> A parolee with a past record of burglaries is
> tracked to a downtown shopping district (in fact,
> is exactly placed in a store known to be locked
> up for the night) and the physiological data
> reveals an increased respiration rate, a tension
> in the musulature and an increased flow of ad-
> renalin. It would be a safe guess, certainly,
> that he was up to no good. The computer in this
> case, weighing the possibilities, would come to
> a decision and alert the police or parole officer
> so that they could hasten to the scene; or, if the
> subject were equipped with an implanted radiotele-
> meter, it would transmit an electrical signal
> which could block further action by the subject
> by causing him to forget or abandon his project.

It does not seem likely that Ingraham and Smith's suggestion would be approved, but the fact that they could propose such a possibility testifies to the impact of the many
unrealistic and distorted representations of brain control
techniques.

PSYCHOSURGERY: TUNNEL VISION

An evaluation of psychosurgical procedures is clearly
beyond the scope of this chapter. The topic is charged
with emotion and sufficiently complex so that anyone wanting
to justify either extreme position would have no difficulty
finding arguments and evidence to support his bias. There
is no doubt that there was evidence in the past and there is
some in the present to support the position that psychosurgery can produce serious deficits. It is also true that

149

the majority of the large-scale studies have concluded
that a significant number of psychiatric patients receiving
psychosurgery have been improved. Almost all of the older
studies can be criticized on various methodological grounds,
such as the involvement of the evaluators with the success
of the surgery. Also, the testing instruments were probably insensitive to important changes in personality and
intellect. Often the estimates of improvement gave too
much weight to the elimination of "troublesome behavior"
and placed much less emphasis on the qualitative level of
the postoperative adjustment. Much of this past information is irrelevant, however, as the patients selected over
the years for psychosurgery have changed and the surgical
procedures themselves have been greatly refined. Only during the past few years have data applicable to current psychosurgical practice started to accumulate. In view of
many past mistakes it seems appropriate to underscore the
point that any evaluation of surgery on any brain structure must be guided by a conviction that postoperative
changes cannot possibly be limited to a narrow behavior
category.

   In the past, and to some extent in the present, a number
of psychosurgeons have displayed a frightening tendency
toward tunnel vision. They have been quick to see possible
clinical applicability in some of the results of studies of
brain-lesioned animals. Unfortunately, they have often
seemed oblivious to behavioral changes that should have
cautioned them against the operation, or at least influenced the procedures used for evaluating the patients post-
operatively. As is true in the field of brain stimulation,
there are many instances where the outcome of a particular
brain operation has been considered only in terms of its
potential effect on one specific behavioral tendency.

   A few brief examples may help to illustrate this point.
It is generally acknowledged that Egas Moniz, the Portu-
guese neurologist who received the Nobel Prize for initiating the prefrontal lobotomy, was greatly influenced by
Carlyle Jacobsen's report of a chimpanzee that stopped
throwing temper tantrums after destruction of its frontal
lobes. The main point of Jacobsen's findings, however,
seemed not to have influenced Moniz. Jacobsen had shown
that monkeys and chimpanzees appear to permanently lose
the capacity to solve certain types of problems following
destruction of their frontal lobes. Moniz neither com-
mented on this fact or attempted to test his patients on
relevant problems. Even though it may prove quite difficult to design appropriate ways to test humans for some of

the deficits displayed by animals, there is no longer any justification for ignoring possible losses in capacities.

One of the more striking illustrations of tunnel vision comes from a psychosurgical procedure used to treat pedophilic homosexuals. These are men who seek out sexual opportunities with young boys. Fritz Roeder and his colleagues at the University of Gottingen have described their response to a film depicting the hypersexual behavior of cats after destruction of a brain region in the temporal lobes called the amygdala nucleus.[12,13] They have written:

> ...the behavior of male cats with lesions of the amygdala region in some respects closely approached that of human perversion. The films convinced us that there was a basis for a therapeutic, stereotaxic approach to this problem in man.

Roeder and his colleagues were referring to the fact that there were reports that ventromedial hypothalamic lesions eliminate the hypersexuality previously produced by destruction of the amygdala. They proceeded to make stereotaxic lesions in the ventromedial hypothalamic nucleus in man. Based on experience with a relatively small patient population that has been described in only a cursory manner in the literature, they presented the following, rather shocking, conclusion about their surgical procedure:

> ...there is no doubt that experimental behavioral research has afforded us a basic method to eliminate or to control pedophilic homosexuality by means of an effective psychosurgical operation in the area of the *sex behavior center*. (emphasis mine)

The reports from this group seem to be amazingly uninfluenced by the voluminous literature which implicates the ventromedial hypothalamus in the regulation of hormonal balance, appetite, irritability, and other behavioral and physiological functions. Apparently once the focus was directed at sexual behavior changes the other behavior changes known to be regulated by the same brain structure were neglected.

Still another example of tunnel vision (from the many that could be offered) can be seen in a report by Flemming Quaade of Denmark.[14] Quaade described the destruction of

151

another part of the brain, the lateral hypothalamus, as a surgical treatment for very obese patients. This operation was undertaken because of many reports that animals show marked decreases in food consumption after this brain region is destroyed. Once again, a large literature was apparently overlooked. There is a considerable body of experimental evidence which indicates that destruction of the lateral hypothalamus reduces general motivation as well as appetite for food. Animals show a significantly dampened responsiveness to all stimuli and impairment in learning capacity.

It is certainly not implied that these examples of tunnel vision are representative of all psychosurgeons. Nevertheless, the dangers of adopting an oversimplified view of brain-behavior relationships can be great. Even a few instances are too many.

PSYCHOSURGERY: INDIVIDUAL VERSUS SOCIAL THERAPY

It is important that psychosurgery as an attempt to help intractably ill mental patients be distinguished from claims that it may contribute significantly to the solution of social problems. This latter possibility has been suggested by a number of people, but most notably by Vernon Mark and Frank Ervin. In their book, VIOLENCE AND THE BRAIN[15], Mark and Ervin stressed the magnitude of violence in the United States and clearly implied that biological interventions can make a significant contribution toward a solution of that problem. The book contains perfunctory concessions to the possible contribution of environmental factors, but these are more than counterbalanced by statements which criticize (often unfairly) the inability of such factors to explain the occurrence of violence, or to be useful in generating helpful programs. By exaggerating the ineffectiveness of environmental programs which have not been tried in any serious way, the biological solutions they offer appear to gain in attractiveness. Mark and Ervin make it very clear in the preface of their book that they "have written this book to stimulate a new and biologically oriented approach to the problem of human violence." They remind us of the severity of the problem by indicating that:

In 1968 more Americans were the victims of murder and aggravated assault in the United States than were killed and wounded in the seven-and-one-half years of the Vietnam War; and altogether almost half a million of us were the victims of homicide,

152

rape, and assault.

The impression that brain abnormalities may be a major cause of violence is conveyed early in the book. Thus Mark and Ervin write:

Most people consider brain disease to be a rare phenomenon. It is likely, however, that more than ten million Americans suffer from an obvious brain disease, and the brains of another five million have been subtly damaged.

Because the statistics on brain abnormalities are so intimately coupled with descriptions of violent crimes, only the most critical reader will fail to draw the conclusion that brain pathology is a frequent cause of violent behavior.

Mark and Ervin suggest that sudden, "episodic," violent behavior is often triggered by abnormal foci in a temporal lobe brain structure called the amygdala. Much of their argument is based on the belief that there is a strong relationship between violent behavior and temporal lobe epilepsy and even "convulsive disorders" without seizures. While there is some evidence to support this view, most current reviews of the literature have concluded that the relationship between temporal lobe epilepsy and violence has been exaggerated.[16] Whatever the true extent of the relationship, there is little justification for the position that a significant amount of the total violence in our society can be attributed to brain damage. Nor can the increases in violence be clearly related to increases in brain pathology.

Because the rationale justifying the brain surgery proposed by Mark and Ervin may be very compelling and their technique appears most elegant, it is especially important to consider the underlying logic carefully. At the heart of their proposal is the claim that electrical stimulation is a reliable means of locating the exact focus in the brain that is presumed to trigger the violent episodes of many persons exhibiting such behavior. This is said to be so because electrical stimulation triggers assaultive behavior only at some brain foci. Once the critical focus is located, they claim it can be destroyed with such precision that only a small brain area need be damaged. This area is likely to have been pathological anyway. Thus they have written:

...tiny electrodes are implanted in the brain
and used to destroy a very small number of
cells in a precisely determined manner.

and also:

To our knowledge, this is the first time
that rage behavior was artificially produced
by electrical stimulation in an abnormal
brain and used to diagnose the proper placement for a therapeutic lesion.

It is important to look more closely at the claim that
electrical stimulation is a reliable technique for determining the brain locus responsible for the behavior.

The experimental and clinical evidence clearly indicates that the behavior evoked by brain stimulation in
animals and man is greatly influenced by the temperament
or personality of the subject, and the general emotional
state aroused by the stimulation. If the brain stimulation evokes fear, anxiety, or pain and the subject is prone
toward violence or aggression, then such behavior will
probably be expressed in response to stimulation at a great
many brain sites which evoke such emotional states. Many
animal studies support this conclusion. Panksepp[17], for
example, has concluded from a study in which brain stimulation provoked rats to kill mice that the ability to evoke
this behavior:

...interacted with the behavioral typology of
individual animals - animals normally inclined
to kill mice were more likely to kill during hypothalamic stimulation than non-killers. Thus,
the electrically elicited response was probably
not determined by specific functions of the tissue under the electrode but by the personality of
the rat.

There is little justification for the belief that brain
stimulation is a valid technique for locating discrete
foci that trigger violence, even if such foci exist.[18]
In the assaultive patients described by Mark Ervin, it
is likely that violence can be triggered by stimulating
a great number of brain sites. It can probably also be
triggered by hard pinching of the skin! It is clear that
the ability of brain stimulation to ferret out a presumed
critical focus of a behavioral trait has been grossly
exaggerated. Moreover, the statement that only "a very

small number of cells" are destroyed in this psychosurgical procedure is not supported by their own reports. Most of the operations destroy parts of the amygdala on both sides of the brain. The general practice has been to gradually increase the size of the lesion over a number of months until the troublesome violent behavior appears to have significantly diminished. Although Mark and Ervin have not been very complete in the description of their surgical procedure, the information they have published strongly suggests that the area destroyed is much larger and less precisely defined than they have implied in their book.

An additional point that should be made is that VIOLENCE AND THE BRAIN conveyed the impression to many readers that the surgical techniques described are applicable to a great number of violent persons, including those who do not have clear evidence of temporal lobe epilepsy. However, Mark and Ervin's own surgical population seems to have been restricted to epileptic patients, or those with reasonably good evidence of brain pathology. In the court proceeding of the well-publicized John Doe case in Detroit, evidence was presented to indicate that the neurologist, Ernst Rodin, had been mislead by the book into believing that the surgical technique had very broad applicability to behavior problems with no demonstrable brain pathology.[19] Indeed, there seems to be little doubt that many other readers concluded that Mark and Ervin were suggesting that the behavior which characterized their "dyscontrol syndrome," might be sufficient grounds alone to justify surgery. It is only fair to note that, whatever his conviction prior to all the criticism of VIOLENCE AND THE BRAIN, Vernon Mark has made it very clear recently that he now believes the neurosurgical treatment of violence should be used very cautiously, and only where the evidence of brain pathology is convincing.[20]

It is especially important that we remain alert to the possibility that the frustration over our inability to stop the spread of violence will cause some people to lose perspective on the problem. It should be evident that humans can exhibit great differences in behavior depending upon their social and physical environment. This is clearly not due primarily to biological difference between people because the same people behave very differently under different conditions. The point is well illustrated in Colin Turnbull's[21] account of the transformation of the Ik. They were a friendly African people forced to change from a life of successful hunting to that of farming under

the most difficult circumstances. Within three generations their social organization had deteriorated. The
struggle for individual survival made them hostile and
as "generally mean as any people could be."

Although it may not be easy to introduce appropriate
social changes to decrease violent behavior, there is
certainly no reason to believe that biological solutions
will soon be available. There will continue to be a few
assaultive patients with clear brain pathology who may
be significantly improved by surgical intervention. However, for the foreseeable future we should increase,
rather than decrease, attempts to find social solutions
for what are primarily social problems.

BRIEF REFLECTIONS ON ETHICAL ISSUES

Over the past several years there has been a great
amount of discussion of ethical issues raised by the new
techniques for physically intervening into the brain.
There has been so much discussion, in fact, that it is
a challenge to find anything new to say on the topic. It
is clear that many of the concerns involve problems that
are not unique to psychosurgery, but are common to most
areas of biomedical experimentation.[22] It is often argued,
however, that operations on the brain present special
ethical problems. These concerns center around the irreversibility of brain damage, because the brain is the
organ responsible for our creativity, individuality and
all that permits us to be human. While there is some
merit to this argument, it is frequently phrased in much
too general a form to be meaningful or guide our ethical
decisions. If it is meant only to imply that we should
be very hesitant to operate on the brain the point is so
obvious that it is almost trivial. Often, however, the
argument takes the form of implying that any brain destruction must produce devastating consequences for intellect, personality, and capacity for experiencing emotions.
This is definitely not true. People who have undergone
relatively uncontroversial brain operations, such as those
to remove tumors or to repair cerebral vascular damage,
do not necessarily experience any great deficits in spite
of brain damage. The results in any particular case depend on the area of the brain damaged and other factors.
It is not at all helpful, therefore, to list all the human
characteristics dependent on the whole brain. We must
come to grips with the more meaningful problem of predicting the likely consequences of specific brain interventions
on particular types of patients. This information should

be evaluated against the background of the alternative therapeutic possibilities available to the patient.

There is a tendency these days to suggest a moratorium to review the evidence whenever there is a strong disagreement about some medical procedure. On the surface, no suggestion could seem more reasonable and moderate. There is no doubt that there are situations where a review of past experience and practice would be most helpful in reaching the right conclusion. It is questionable, however, whether psychosurgery is such an area. The data necessary for evaluating the newer psychosurgical procedures is just beginning to accumulate. It is unlikely that the evidence presently available will make it possible to reach a consensus. We are likely to remain frozen in a moratorium status while the more extreme protagonists continue to cite only that part of the inconclusive evidence which supports their views. A moratorium is most useful in those instances where it is reasonable to expect that an objective review of the available data will provide answers to the questions raised. It should not be used as a device to obtain a *de facto* prohibition. Psychosurgery aside, the fact that there have been instances in the past where patients have been deprived of effective treatment because of misguided objections should caution us against overuse of the moratorium solution.

Rejecting the usefulness of a moratorium on psychosurgery does not imply that anarchy should prevail. The recognition that a medical procedure is experimental implies the need for control. The process of reviewing proposed experimental procedures is clearly at the very center of the problem of patient protection. Traditionally, review panels in hospitals have consisted of physicians reviewing the proposals of their colleagues from the same institutions. This procedure has not worked very effectively in the past for many reasons. Physicians, particularly those at the same institution, have often been very reluctant to criticize their colleagues. Moreover, review panel members are often too busy to critically evaluate the relevant literature or the adequacy of previous attempts to treat the patient by less "heroic" procedures. The solution to this problem is not easy. It seems that people willing to contribute adequate time to serve on review panels often have strong biases that color their judgment. Some method must be established, perhaps similar to the jury selection system, to obtain reasonably objective people to serve on review panels.

The question of who should serve on review panels is quite controversial. Many physicians feel quite strongly that the layman has no place on such panels. For example, M. Hunter Brown, a neurosurgeon practicing psychosurgery in California, has clearly stated his views on the subject:

Current proposals for surgical boards manned by lawyers, ethicists, consumer advocates, clergymen, etc., are arrant nonsense; lay persons would get into a scientific act in which they have no competence. An informed decision on target neurosurgery is not possible for uninformed people, particularly from those not acquainted with modern technology.[23]

Not everyone is persuaded by such arguments. Clearly, the present review practices have often not provided adequate protection for the patient. Moreover, it is a mistake to believe that the only responsibility of review panels is to consider technical aspects of proposals. Patient protection involves decisions on questions about the adequacy of the consent obtained from the patient or guardian, the thoroughness of previous attempts to treat the patient by more conservative therapies, legal questions about the patient's status, and a number of other problem areas that can be dealt with most adequately by a heterogeneous review panel. It is also likely to be a more effective means of monitoring the rare, but real, problem of the irresponsible physician, or one whose ego involvement in a new procedure may interfere with judgment. Heterogeneous review panels may be better suited to help in the systematic pooling of objective data that would be most useful in estimating the probability of future success.

It is clear that we must find a way to institute whatever changes are necessary to assure that members of review panels take their job seriously. During the past few years there have been several reports of patients being subjected to very questionable experimental procedures. Unfortunately, no one has asked the review committees in these instances to justify their approval. In the John Doe case in Detroit, a case involving a proposal to perform an amygdalectomy on an incarcerated patient, one of the review panel members failed to attend any of the meetings related to the proposed surgery. It was his view that:

As a layman I am unqualified to comment on any of

158

the technical aspects which are involved in the
project. Therefore, we must all trust the good
intentions and technical competence of the Hospital Medical Committee, psychologists, psychiatrists, neurologists, etc., who have reviewed
and evaluated John Doe's case.

Such passivity, whether from laymen or physicians, is
totally misplaced on review panels. Although every member of a review panel should hold the patient's best
interest above all else, it might be useful if one person
was appointed the patient's ombudsman. The ombudsman
would be responsible for making certain that the information presented to the panel was as complete and objective
as possible.

One recurrent problem concerns the ambiguity about
which procedures are experimental. A given procedure
may require a thorough review in one hospital because it
is considered experimental. In another hospital the same
technique, for example, implantation of brain stimulating
electrodes, may be classified as a diagnostic tool and
may not be reviewed at all or it may receive only a
cursory review. It is obvious that some classification
standardization would help protect patients. There are
other reasons for differences between hospitals. The control over experimental medicine has commonly been regulated
by the Department of Health, Education, and Welfare's
prerogative to withhold federal funds. Some private hospitals, which receive no such funding, are relatively
uncontrolled. Under the present system, a patient at a
hospital receiving no federal funds may be much more vulnerable than one at a hospital required by HEW to utilize
an approved review procedure.

Briefly we need to establish standards for classifying
procedures as experimental. Review panels must not
serve as rubber stamp committees. The members must be
made to realize that they may have to justify their decision. There may actually be a legal precedent here;
review panel members have been sued because of the claim
that their decision was clearly not based on the best
interest of the patient. Panels should have multidimen-
sional representation, but if this is going to work some
mechanism must be established for selecting reasonably
unbiased members. Lastly, since the procedures in question are experimental there must be adequate assessment
procedures established beforehand. Mechanisms for col-
lecting, pooling, and disseminating information on the

159

outcome must also be developed.

THE ETHICAL BEHAVIOR OF ETHICISTS

Evidence of the increasing concern over the ethical and social implications of scientific developments is available almost everywhere. Symposia on ethical issues are now on the programs of meetings in almost all of the sciences and professions. This is a very healthy development, but is not insulated from the influence that can distort every human endeavor in our society. While the motivation of most people concerned about the ethical issues may be completely admirable, it would be a mistake to believe that all the good motives belong with those who espouse ethical causes and all the bad motives reside with those they criticize. Ethics is a rapidly expanding profession and as a result it would be unrealistic to believe that its members are any more immune to the lure of prestige, power, position, and prosperity than are those engaged in research. Motives aside, the important issue is that criticisms based on ethical concerns can have a great impact on biomedical research and public policy. Often this impact is not adequately considered. Usually the researcher and the clinician are put on the defensive while the criticism itself is seldom scrutinized. An example may help to illustrate the point.

Concerned individuals, probably with motives that could not be impugned, argued passionately for indeterminate jail sentences to remove the sentencing power from vindictive judges and as a way to release prisoners earlier if they reformed. Much evidence was brought forth to illustrate the capriciousness of some judges who consistently gave very long sentences. There was much less discussion, however, about the practicality of the solution. Most states passed resolutions supporting the indeterminate sentence. The result, which might have been foreseen if more thought was given to the factors influencing decisions about releasing prisoners, is that prisoners receiving indeterminate sentences usually serve longer sentences than those who receive fixed sentences for the same crime.[24] Moreover, they serve their time under the added psychological strain of uncertainty about when they will be released.

Obviously, it is not sufficient to have your "heart in the right place." The harm that is caused unintentionally still hurts. Because concern with ethical issues and reforms does not guarantee beneficial results it is most important to apply the same criteria to those who point out

160

abuses and advocate reform as we do to those engaged in the practices being scrutinized. If we did this, we would find many instances of failure to meet these criteria, particularly when those concerned become convinced they are engaged in a moral crusade.

Psychosurgery is clearly one of the controversial issues that seem to spawn moral crusaders. Some consider this procedure nothing less than a criminal mutilation of the brain. They believe it changes human beings into emotional and intellectual vegetables in order to eliminate their troublesome behavior. Others, however, believe these brain operations are the only means currently available for alleviating certain crippling mental illnesses. The charges and counter-charges are often quite heated.

Although we may understand why emotions enter into some controversies, the intensity of feelings does not justify biased, irresponsible, and even dishonest reporting of the evidence. Although propaganda techniques can be justified under some circumstances, there is no justification for those posing as experts to mislead policy makers by quoting out of context, or reporting only that data which supports a preconceived position. In the long run, the passive acceptance of such distortions as a way of life - or even attempts to justify them on the basis of the goal that is served - will do greater harm than most of the practices that are being criticized.

For several years the Washington, D.C.-based psychiatrist, Peter Breggin, has been one of the leaders in the attack against psychosurgery. It is clearly his right to carry on whatever legal-political action he feels is justified. It should not be his prerogative, however, to pose as an authority while giving distorted testimony to Congressmen or in a courtroom.[25] It is very obvious that Breggin searches the literature only to find ammunition for his cause. Anyone who has objectively examined his statement published in the CONGRESSIONAL RECORD (February 25, vol. 18, no. 26), would have no trouble detecting numerous examples of all the classic ways of distorting evidence. He has selected data, quoted out of context, and resorted to demagogic accusations in order to raise emotions and secure allies. Later, while testifying before the Kennedy Subcommittee, Breggin actually referred to this distorted document as his "reasearch papers in the Congressional Records" (Hearings 359). In the CONGRESSIONAL RECORD Breggin accused O.J. Andy, a Mississippi neurosurgeon, of concealing the fact that he was operating mainly on blacks. This charge was picked up by many persons

161

and repeated frequently. EBONY magazine, for example, published an article entitled, "Psychosurgery: A New Threat to Blacks." Apparently no one bothered to check the facts. According to a letter written in response to my inquiry, Andy claimed that of the forty psychosurgical operations he had performed, only 5% - that is, two cases - of patients were black. This is very significantly below the percentage of blacks in Mississippi. Hearing Breggin's charge, however, black Congressmen and women became understandably concerned. They were among the leaders in proposing legislation (H.R. 6852) to outlaw psychosurgery. Leaving aside any evaluation of psychosurgery, there are obvious dangers in establishing a precedent of passing hastily written legislation in response to political pressure based on distortions. It would be an especially serious mistake to use the legislative route as a means of resolving controversies over specific medical procedures. This is not to deny that Congress may play a useful role in establishing legal guidelines to protect patients.

Electroconvulsive shock treatment (ECT) is another controversy that is presently generating much heat. John Freidberg, a psychiatric resident in Oregon, has written an article on ECT in PSYCHOLOGY TODAY (August 1975). The article's tone is revealed by its title, "Let's Stop Blasting the Brain," and by such bold-type paragraph headings as "Shock Treatment Burns the Brain," "Beating Up the Insane." The text is characterized by such statements as the following:

Egas Moniz experimented with prefrontal lobotomies. In Rome, Ugo Cerletti developed electroconvulsive shock treatment. The Germans came up with a simple and final solution for mental illness; in the late 1930's 275,000 inmates of German psychiatric institutions were starved, beaten, drugged, and gassed to death.

Among other distortions, Friedberg's article clearly conveys the impression that studies by neuropathologists "consistently show severe brain damage" after electroshock treatment. There is no such evidence. New neuroanatomical techniques might reveal brain pathology, but there is no reliable evidence today that this is the case.

It is not unreasonable to expect scholarship, intellectual honesty, accurate treatment of data, and responsible behavior by those who claim to be arguing the ethical position, as well as by those engaged in the controversial

162

experimentation. It is indeed ironic that in the name of ethics, the study of moral behavior, so many questionable acts of morality have been committed. There is no justification for a condition that forces only the researcher to defend himself while leaving the self-appointed defenders of patients' rights, who often have an equally great impact on patient care, completely uncriticized. Important decisions should not be based on demands, demagogy, and distortions. Nor can complex issues usually be reduced to the absurdly simplistic level of the good guys versus the bad guys. Up to now there has been little criticism directed toward those who attempt to influence and coerce in this fashion because of the superficial belief that they are in the service of the angels. Articles such as Friedberg's, however, have been instrumental in initiating legislation in several states which may prevent some patients from what many (probably most) psychiatrists feel is the best available treatment, whatever its shortcomings, for some types of severe depression.

It is sad, but perhaps true, that confrontation may often be the only way to bring attention to a problem. Those who are unresponsive to more rationally presented arguments must be given much of the blame for this development. Unless ways are found to increase the responsiveness of the decision makers, confrontation is likely to continue to be a way of life. However, the techniques that have become necessary to get problems about biomedical research into the open and on the agenda may be totally inappropriate for their resolution. We need muckerackers, gadflys, and probably also demonstrators, but the task of making policy in these complex areas can only be harmed by those who persist in distorting the evidence. Especially great harm can be done by those who adopt the adversary posture of a moral crusader while posing as an expert witness.

Although many positive results have come from the increased concern over practices in biomedical research, we should be sensitive to problems that may develop in the process. It might be useful to mention a few such problem areas. Up to now they have been relatively neglected. To begin, it is important to recognize that only a few of the so-called ethical problems in biomedical research actually involve disagreement about ethics or morality. Those who disagree about psychosurgery, electroshock treatment, implanted electrodes, amphetamines for hyperkinetic children, and many other controversial therapies all claim to have the best interest of the patient in mind. Much of the

163

controversy really is over the true estimate of risk and possible benefits. The resolution of the controversy, therefore, must depend to a great extent on a realistic appraisal of the evidence. If the evidence is not conclusive it may be appropriate and feasible to obtain the data required to make the best judgment. We should recognize that ethicists[26] do not have any set of agreed upon principles that can help us make the right decision. At the same time, they may serve the very useful purpose of raising a problem and discussing the ramifications of different proposed solutions.

The shortcomings of existing experimental procedures in medicine must be evaluated together with the practical alternatives that exist. This is seldom done. It is much safer, and also easier, to criticize than to offer an alternative program. Those who do not have to face patients can afford the luxury of elaborating on our ignorance about the brain and behavior and the causes of mental illness. They are not confronted by a desperate patient and family. The patient cannot always wait for our knowledge of the brain to increase, or for society to become more just. Those who criticize electroshock treatment, for instance, seldom offer any suggestions about how to treat seriously depressed patients. Often suicidal, they do not respond to drugs, psychotherapy, or any other kind of treatment that is now available. There is a real danger that ethical discussions may become so abstract that the patient is no longer visible.

Finally, it is important to recognize that we must be concerned about protecting future patients as well as those in the present. There is a real danger that we might cheat future patients by creating such a bureaucratic morass to protect patients from every conceivable abuse, that the most imaginative and productive researchers will be driven out of every controversial field. It might do us no harm in this context to bear in mind the International Code of Medical Ethics:

> As a stream cannot rise above its source, so a code cannot change a low-grade man into a high-grade doctor, but it can help a good man to be a better man and a more enlightened doctor. It can quicken and inform a conscience, but not create one. (International Code of Medical Ethics, Adopted by the General Assembly of the World Medical Association, London, 1949)

164

FOOTNOTES

[1]Only a few aspects of this complex subject can be discussed here. I have tried to present a readable and unbiased description of the major historical, scientific and ethical considerations in *Brain Control: A Critical Examination of Brain Stimulation and Psychosurgery* (New York: John Wiley, 1973).

[2]M. Chricton, *The Terminal Man* (New York: Alfred Knopf, 1972).

[3]D. Rorvik, "Someone to Watch Over You (for less than 2 cents a day," *Esquire*, (1969), pp. 72, 164.

[4]P. London, *Behavior Control* (New York: Harper & Row, 1969).

[5]K. Clark, Cited in *APA Monitor*, (October, 1971), Vol. 2, No. 1.

[6]J.M. Van Buren, "Evidence Regarding a More Precise Localization of the Posterior Frontal-caudate Arrest Response in Man," *Journal of Neurosurgery*, (1966), pp. 24, 416-417.

[7]E.S. Valenstein, "Behavior Elicited by Hypothalmic Stimulation. A Pre-potency Hypothesis," *Brain, Behavior and Evolution*, (1969), pp. 2, 295-316.

[8]E.S. Valenstein, *Brain Control: A Critical Examination of Brain Stimulation and Psychosurgery* (New York: John Wiley, 1973).

[9]E.S. Valenstein, V.C. Cox and J.W. Kakolewski, "Reexamination of the Role of the Hypothalamus in Motivation," *Psychol. Review*, (1970), pp. 77, 16-31.

[10]The fear that psychosurgery is primarily a behavior-control technique of potentially wide applicability to political and social problems has had a significant influence on legislation under consideration. For instance, a recently proposed federal law (H.R. 6852), which would outlaw psychosurgery, defined the purpose of these operation as:

"(A) modification or control of thoughts, feelings, or behavior rather than the treatment of a known and diagnosed physical disease of the brain;
(B) modification of normal brain function or normal brain tissue in order to control thoughts, feelings, action, or behavior."

165

The arguments frequently reflect political and social concerns. The psychiatrist Peter Breggin has charged, for example, that "these brain studies are not oriented toward liberation of the patient. They are oriented toward law and order and control, toward protecting society against the so-called radical individual." At a "Forum on Psychosurgery" held in Boston (May 4, 1972) a document was circulated which included such statements as:

> "The Boston Area Medical Challenge Clubs, composed of members and friends of the revolutionary communist Progressive Labor Party, are putting out a statement to explain why we feel that the fight against psychosurgery and the fight against capitalism are integrally related."

> "We see all psychosurgery, including that on apolitically mentally 'ill' people, as a form of social control. We see mental distress as a reaction to the world. We reject the proposition that the source of the distress is inside the person's brain."

The unfounded speculation by Mark, Sweet, and Ervin that the more violent participants in race riots might have had brain pathology has undoubtedly caused much anxiety about future political applications of these techniques. It is only fair to note, however, that their surgical patients have all been white and seem to have had demonstrable evidence of brain pathology.

[11] B.L. Ingraham and G.W. Smith, "The Use of Electronics in the Observation and Control of Human Behavior and its Possible Use in Rehabilitation and Parole," *Issues in Criminology*, (1972), pp. 7, 35-53.

[12] F. Roeder, H. Orthner and D. Muller, "The Sterotaxic Treatment of Psychoses and Neuroses." In W. Unbach, ed., *Special Topics in Sterotaxis* (Stuttgart: Hippodrates-Verlag, 1971), pp. 82-105.

[13] F. Roeder, H. Orthner and D. Muller, "The Sterotaxic treatment of pedophilic homosexuality and other sexual deviations. In E. Hitchcock, L. Laitnen and K. Vaernet, eds., *Psychosurgery* (Springfield, Ill.: Charles C. Thomas, 1972), pp. 87-111.

[14] F. Quaade, "Sterotaxy for Obesity," *Lancet*, (1974), pp. 1, 267.

[15] V.H. Mark and F.R. Ervin, *Violence and the Brain* (New York: Harper & Row, 1970).

[16] M. Goldstein, "Brain Research and Violent Behavior," *Arch. Neurology*, (1974), 30, pp. 1-35.

[17] J. Panksepp, "Agression Elicited by Electrical Stimulation of the Hypothalamus in Albino Rats," *Physiological Behavior*, (1971), 6, pp. 321-329.

[18] E.S. Valenstein, "Brain Stimulation and the Origin of Violent Behavior." In W.L. Smith and A. Kling, eds., *Issues in Brain/Behavior Control* (New York: Halsted Press, 1976).

[19] R.S. Gass, "Kaimowitz vs. Department of Mental Health: The Detroit Psychosurgery Case." In W. Gaylin, J. Meister and R. Neville, eds., *Operating on the Mind. The Psychosurgery Conflict* (New York: Basic Books, 1975), pp. 73-86.

[20] V.H. Mark and R. Neville, "Brain Surgery in Aggressive Epileptics," *JAMA*, (1973), pp. 226, 765-772.

[21] C.M. Turnbull, *The Mountain People* (New York: Simon & Shuster Touchstone, 1972).

[22] The word "experimental" often conjures up the image of human "guenea pigs." The use of the term here does not imply that the primary purpose is research rather than therapy. A procedure may be considered "experimental" if the results are not predictable or if there is no clear agreement on the value of the treatment.

[23] M.H. Brown, *The Captive Patient* in press.

[24] A recent Report, "Time Served, Parole Outcome, and Type of Sentence," (National Council on Crime and Delinquency, Davis, California) indicates that the reverse may be true if the "maximum sentences" are very long.

[25] S.I. Shuman, "The Emotional, Medical and Legal Reasons for the Special Concern About Psychosurgery." In F.J. Ayd, Jr., ed., *Medical, Moral and Legal Issues in Mental Health Care* (Baltimore: Williams & Wilkins, 1974), pp. 48-80.

[26] Included here are also those philosophers specializing in ethics. It is certainly not meant to imply, however, that the behavior of the type of "moral crusader" described earlier

is characteristic of the serious scholars grappling with the difficult problem of systematizing ethical decisions. It has been suggested that the label "ethicist" should be reserved for the latter, but it has frequently been used to refer to any one strongly committed to the solution of ethical problems.

MOVING PEOPLE TOWARD PERFECTION BY MEDICAL TECHNOLOGY

William P. Wilson
Duke University Medical Center

Dr. Valenstein's excellent presentation has thoroughly reviewed the subject of brain control. I wish to express my appreciation for his views because they have added a dimension of sanity to discussion of the subject.

IMPERFECTION AND CONTROL

All around us in the world today is evidence of man's imperfection. Murder, civil war, regional wars, rape, sexual perversions, the rising divorce rate as well as increasing alcohol and drug abuse are expressions of this imperfection. As we begin to focus on the individuals who make up our society, we observe that there are varying degrees of perfection. The more imperfect are readily identifiable, for they fill our jails and mental hospitals. We have to work to find more nearly perfect persons, but they are also there feeding the hungry, giving drink to the thirsty, receiving strangers in their homes, clothing the naked, taking care of the sick and visiting the prisoners. One doesn't get a headline for doing these things. There are no laws against perfection (Matt. 25:35-36, Gal. 5:23), so no one pays attention to the nearly perfect man.

It is unfortunate that we cannot identify our nearly perfect population because this leads us to believe that man's capacity for imperfection is greater than his capacity

WILLIAM P. WILSON is Professor and Head of the Division of Biological Psychiatry at the Duke University Medical Center in Durham, North Carolina. Dr. Wilson has edited one book and published over 140 articles in scientific periodicals, a number of them focusing on the technical and evaluative aspects of psychosurgery. He is an examiner for the American Board of Psychiatry and Neurology, a diplomate of the National Board of Medical Examiners, and a member of the Governor's Committee, Task Force on Diagnosis and Treatment, and President of the Southern Psychiatric Association.

for perfection. He was created with the capacity for both and he knows it, but he doesn't choose to seek perfection. Because he does not seek perfection God has given him laws to show him that he has imperfection, and punishment to cause him to strive to avoid behaving imperfectly. In a like manner God has commanded him to seek perfection (Matt. 5:48), in order that he may live in harmony with his fellowman and with God.[1-5] In this regard, I define perfection as walking according to the ways of God by His Spirit's empowering.

But how do human beings come to manifest imperfection so readily? Careful inspection of the origins of human behavior reveals to us that man has either learned these imperfections or that he has acquired them in genetically determined defects and diseases and from trauma, infection and neoplasm.

Granting that man has some imperfections for which he is not responsible, what are we to do about them? Do we have a responsibility? The answer to this question is "yes." We are told to heal the sick (Matt. 10:8). Obeying this and the order to preach the Gospel, Christians are to strive to heal both man's learned and inherited imperfections. We are to help him approach the highest level of perfection possible.

For thousands of years those in medicine, science, and theology have tried to heal man. However, man's sinful nature has caused him to put himself above the law and allow his imperfection to abuse the entrusted capacity God gave him to be a helper. He has used knowledge for his own ends and not for the good of mankind. One aspect of his desire for disobedience has been the notion that he might develop methods of controlling his fellow man's behavior, such as behavior control. In the past he used force but more recently he has attempted to do this by psychological techniques[6] or by altering the physiological activity of the brain.[7]

The mechanisms that mediate our repertoire of behaviors are found within our brain. The most basic of these are a spirit or life force and a soul made up of our intellect and emotional responses. These mechanisms drive man to maintain and even extend this integrity beyond the grave. They also drive him to defend his integrity and finally to recreate himself. The complex behaviors that grow out of these drives are useful imperfections. We as a nation used one of these qualities most commonly considered as imperfections (aggression) to gain our freedom from Britain. We used this same one to overcome the evil incarnate in Adolph Hitler. There are many other examples. They are a necessary part of our survival behaviors. We should not want to destroy man in the process of eradicating those behaviors that are potentially misdirected. It really is not the possession of these behaviors but the lack of control that has proven so deleterious. We have tried to maintain control of man with the law, but law quite clearly cannot do it (Rom. 3: 19-22).

If the law cannot do it, how in the world can we hope to control man? I am not certain that we would want to do so using the technology being discussed here. Because of man's capacity for imperfection there are those who want to use these techniques for their own ends. They would not put them to beneficial use but instead would use them for evil ends. It is no wonder that people get frightened when men like Delgado, Ervin or others[8] talk as they do about brain or mind control. The layman has witnessed man's rebellion and the fruit of that rebellion - man's inhumanity to man and it is no wonder that they suspect evil intentions.

TRUTH AND THE LAW

On the other hand, there are many who desire to make their brother whole, healing and helping him to become more perfect. Should we do so, even when there are those who accuse us of evil intent? The answer, I believe, is yes. It is still a fact that a person who acquired behavioral imperfections through disease can be investigated and treated with the techniques under discussion. As of now stimulation, psychosurgery, electroshock and psychopharmacology are the only tools that we have to do this with. These tools have been given to us (indirectly, of course) by God. We believe this for all knowledge comes from God. Since He wishes us to compassionately use our knowledge to extend our healing power we have no choice but to do this. We must, therefore, seek the truth and utilize this truth to deal with man's imperfections. We are to be investigators!

Investigative efforts should be designed as a means by which we carry out God's commands to take dominion over the earth. This command was partly heeded when Abraham left Ur and went to Israel. In modern life, man has also explored the new world, the moon, and I am sure will eventually explore the planets. At the same time that he extended his horizons outward, he also turned them inward and began to investigate the microscopic and submicroscopic world. He

171

has investigated his body, his mind, compounds, molecules, and atoms.

Because of the extraordinary mystery of the mind-brain relationships, attention has been particularly focused on the brain and attempts to decipher the chemistry of behavior, as well as the physiological relationships of the brain to behavior. Of late these efforts have been hampered by the negative emotional cries of a few who are afraid of control. They see themselves being turned like a "Brave Bull," or visualize themselves walking about like Zombies "spaced" on some mind-controlling drug.[9] The outgrowth of this fear is litigation and threats of litigation that leave medicine immobilized. One and a half million dollars is a lot of money; none of us can afford to be involved in such a risky business. Therefore, medical scientists wonder why they should use the technology. But this creates guilt so we rationalize that man does not deserve anything better than what he already possesses. At that point we have become the victims of our own lack of control over irresponsible investigators and therapists. Among them are those who advocate the use of fetuses for non-survival experiments, those who do psychosurgical procedures using the most primitive techniques, those who administer drugs in mind boggling amounts and those whose desire for glory cause them to make foolish statements that increase the fears of the less informed laity.

Christians must take the lead in denouncing this kind of irresponsibility. We must seek and present truth to those who make laws so that our suffering brothers and sisters will not be deprived of treatment that would help them. Controls are necessary and must be available to protect mankind from irresponsible scientists. However, those controls should not deprive him of the right to treatment.

If we possess truth we should put it to use. I believe that man has free will as long as his mind is normally functioning. The law is of the same persuasion. It has, therefore, provided legal safeguards to protect those who do not have a normally functioning mind. Man has alternatives when he is mentally normal. When he is not capable of choosing alternatives, those who love him most should have the freedom to choose for him. Unfortunately, both scientists and governments often make the choices for even the normal man. They do so by due process of law or despotically when the general public feels incapable of using their decision making rights.

Ultimately, though, someone does have to make a decision. How do we decide when and under what circumstances the procedures mentioned should be applied? Medicine has normally made decisions using a model in which the positive (healing) aspects of any kind of intervention that involves exploration of man's inner space are weighed against the damage likely to be done. *Primum non nocere* is our admonition, so we must consider carefully whether we will do harm that will make the patient even more incomplete. This is especially true in regard to his brain. To make any kind of judgment we must weigh the possibilities of helping the patient against the possibility of causing him further losses.[10] Let me illustrate from my own career.

Some years ago Dr. Blaine Nashold and I began the investigation of chronic central pain.[11] Our investigations[12-16] were stimulated by our desire to help a woman who for eight years had excruciating pain of her face and shoulder. This had developed after a stroke. In planning her evaluation for treatment we had to consider the following indications for investigation:

1. There was no treatment that relieved her pain adequately;
2. All we could do was to investigate her with stereotractically implanted electrodes in critical brain structures. Our rationale was based on the suspicion that these structures were hyperexcitable;
3. We were familiar with the technology, having used it in epileptics for fourteen years. We knew it was safe;
4. No other techniques were available that offered any hope of obtaining information that might eventually result in relief;
5. If we did nothing the patient would continue to suffer.

After carefully explaining to her and her husband what we were about, and obtaining their consent, we went ahead with the investigation. We learned that: the pain came from hyperexcitable neurons in the mesencephalon; a small radiofrequency lesion in her most active epileptic area essentially relieved her of pain; a possible new understanding of pain perception was gained; stimulation and recording studies conducted gave us new understanding of the physiology of emotion and sleep; no harmful effects resulted from the implantation of electrodes in deep structures, though the lesion resulted in some defects of eye movement.

173

Let us take an example from psychosurgery.[17] In 1949, the author was appointed the assistant director of a large psychosurgery program (approximetely 400 procedures). The surgical procedure used was the one described by Lyerly and Poppen. It was done by a competent neurosurgeon. The program was begun because there were 8,000 patients in three state institutions, at least 50 percent of whom were chronic schizophrenics. Electroshock and insulin coma treatment, the only treatments available, had been unsuccessful in relieving all of those patients' illness. Since most of these patients had been hospitalized eight years or longer it seemed unlikely that spontaneous recovery would occur, and our examination led us to believe that they were likely to live out their lives in a state mental hospital.

We believe the lobotomy was technically indicated because previous investigators had reported that sixty-six percent of the patients would be improved and that thirty-three percent would be unchanged. Of those improved, thirty-three percent would be able to leave the hospital. The technique was applicable because the mortality rate with good surgery would be negligible, complications would be limited to occasional seizures in a few patients, and selection was based on the opinion of three competent psychiatrists who were in agreement that the procedure was indicated.

The outcome of our specific program was that fifty percent of the patients were discharged from the hospital, and twenty-eight percent became self-supporting. In addition, the quality of life of those not discharged was improved or unchanged. Seizures occurred in 0.1 percent of the patients and were easily controlled with minimal doses of dilantin. No adverse results were observed in the 400 patients carefully followed. Two hundred and fifty of them were followed for fifteen years.

Physicians and psychiatrists have normally utilized medical technology to fulfill the requirements of the laws of beneficence. I feel in these two instances that these laws were obeyed.[18]

THE FUTURE OF BRAIN CONTROL

I would like now to turn my attention to the specific technologies that are under consideration. My discussion here is to establish my right to speak from experience, for I have used the technologies under discussion for twenty-seven years.[19-30] I have stimulated the brain of 100 patients an average of six times in approximately 36 locations.

Each patient was tested twice and approximately 5,000 stimulations have been performed. I have personally been
involved in the selection, operation, and follow-up of 400
patients treated using a psychosurgical procedure, and have
administered or supervised the administration of nearly
60,000 electroshock treatments. Finally, I have extensively
used or conducted investigations on all of the major neu-
ropharmacologic agents as well as a few hallucinogens.

On the basis of this experience I agree with Dr.
Valenstein's opinion that I cannot foresee man's technology
becoming so refined that he can plug into the intricate circuitry of the brain and take control of behavior. I would
argue that it is unlikely that we will ever be able to
technically provide the stimulation that will allow us to
influence significantly and extraordinarily complex events
occurring in even the simplest of brain systems.

I also have to say that no other technology including
neuropharmacology currently offers a possibility of useful
control of a person or population.

My reasons are summarized as follows:

1. All destructive technologies decrease perfection in
some way
2. All excitatory technologies unbalance homeostasis and
as a result cannot be well controlled. They tend to
trigger massive inhibitory responses which inactivate
many systems. To date, none elicit useful behavior that
could be used for malevolent ends
3. Many drugs that influence behavior produce their effect by normalizing excessive excitation or inhibition
occurring as a result of disease. They are not 100 percent effective. Most have no useful effects on persons
who are not diseased. Drugs that excite or depress cause
too many other undesirable effects.

CONTROLLING THE TECHNOLOGY

Finally, we come to the issue of control of technology.
In general, medicine, because of its ethics, has been given
control of its technologies. Medicine, however, was dis-
credited after World War II because of its unethical activities in Nazi Germany and more recently in Communist Russia.
The behaviors of some persons in medicine in our country
have caused increasing suspicion that physicians might not
be trustworthy. The result has been the introduction of an
increasing amount of legislation which is aimed at control.

175

In essence the laws have decreased the use of beneficial techniques such as electroshock, psychosurgery, and investigative efforts that are aimed at determining the origins of mental functioning. Patients are being deprived of treatment and/or released from hospitals only to be returned to jails and skid rows by those who claim they are protecting people from damaging treatment. Instead they are condemning many of them to the complications of pharmacotherapy (dyskinesias and other central damage), or to live in a hostile world where they will be subject to human vultures who prey on those whose mental disability leaves them defenseless. The reformers really don't mean to do this. They simply have tunnel vision and cannot see the entire picture. My hope is that responsible scientists can be agents to urge the presentation of truth. We can then pray and work for responsible legislation that places appropriate controls on our irresponsible colleagues. At the same time legislation must protect the right to treatment of our imperfect citizens who desire to be more perfect.

Mankind has the right to be relieved of suffering, have disease treated, and to have new treatments developed that will offer him more hope.

FOOTNOTES

[1]L. Von Bertalanffy, "Chance or Law," in A. Koestler and J. R. Smythies, *Beyond Reductionism* (New York: MacMillian Co., 1969), pp. 56-84.

[2]L.P.Bird, "TAO Principles of Medical Ethics," in C.A. Frazier, *Is it Moral to Modify Man?* (Springfield, Ill.: C. C. Thomas, 1973).

[3]Claude A. Frazier, ed. *Is it Moral to Modify Man?* (Springfield, Ill.: C.C. Thomas, 1973).

[4]William James, *The Varieties of Religious Experience* (New York: Collier Books, 1961).

[5]P.A. Weiss, "The Living System: Determinism Stratified," in A. Koestler and J.R. Smythies, *op. cit.*, pp. 3-55.

[6]Perry London, *Behavior Control* (New York: Harper & Row, 1969).

176

[7] Elliot S. Valenstein, *Brain Control: A Critical Examination of Brain Stimulation and Psychosurgery* (New York: John Wiley & Sons, 1973).

[8] W.C.A. Sternbergh, W.P. Wilson and J.L. Hughes, "Anticholinergic Hallucinosis. II. Effects of Atropine and JB-329 on Activity of the Visual System, Nonspecific Projection System, and Hippocampus in Animals with and without Reticular System Lesions," *Recent Advances in Biological Psychiatry*. Vol. 8, Plenum Press, (1966), pp. 187-197.

[9] Elliot S. Valenstein, *op. cit.*

[10] L.P. Bird, *op. cit.*

[11] B.S. Nashold and W.P. Wilson, "Central Pain, Observations in Man with Chronic Implanted Electrodes in the Midbrain Tegmentum," *Confinia Neurologica*. 27: 30-44, (1966).

[12] B.S. Nashold, J.P. Gills and W.P. Wilson, "Ocular Signs of Brain Stimulation in the Human," *Confinia Neurologica*. 29: 169-174, (1967).

[13] B.S. Nashold and W.P. Wilson, "Thalamic and Mesencephalic Epilepsy in the Human," *Revista del Instituto Nacional de Neurologia*. No. 4, 1: 22-31, (June, 1967).

[14] B.S. Nashold, W.P. Wilson and D.G. Slaughter, "Stereotaxic Midbrain Lesions for Central Dysesthesia and Phantom Pain," *Journal of Neurosurgery*. 30: 116-126, No. 2. (1969).

[15] B.S. Nashold, W.P. Wilson and D.G. Slaughter, "Sensations Evoked by Stimulation in the Midbrain of Man," *Journal of Neurosurgery*. 30: 14-24, (1969).

[16] B.S. Nashold and W.P. Wilson, "Central Pain and the Irritable Midbrain," *Pain and Suffering*. Benjamin L. Crue, ed., (Springfield Ill.: C.C. Thomas, 1970).

[17] L.B. Hohman, G.L. Odom, R.B. Suitt, W.P. Wilson, R.C. Carroll, S.E. Glass and J.P. Murdoch, "Preliminary Report on Follow-up of Prefrontal Lobotomies Performed by the Duke Neurosurgical Staff," *North Carolina Medical Journal*. 12: 529-534, (1951).

[18] L.P. Bird, *op. cit.*

[19] Stephen C. Boone, B.S. Nashold Jr. and W.P. Wilson, "The Effects of Cerebellar Stimulation of the Averaged Sensory Evoked Responses in the Cat," Reprinted from: *The Cerebellum, Epilepsy and Behavior*, ed. By Irving S. Cooper, Manuel Riklan and Ray S. Snider, Plenum Publishing Co.

[20] K.C. Fischer and W.P. Wilson, "Methylphenidate and the Hyperkinetic State," *Diseases of the Nervous System*. 32: 695-698, (1971).

[21] J.S. Glotfelty and W.P. Wilson, "Effects of Tranquilizing Drugs on Reticular System Activity in Man," *North Carolina Medical Journal*. 17: 401-405, (1956).

[22] W.W. Spradlin, W.C.A. Sternbergh, W.P. Wilson and J.L. Hughes, "Anticholinergic Hallucinosis. I Effect of Atropine and JB-329 on 'Caudate Spindle' Phenomena and Electrical Activity of Cat Hippocampus," *Recent Advances in Biological Psychiatry*. Vol. 8, Plenum Press, (1966), pp. 175-185.

[23] W.P. Wilson, L.B. Hohman and G.L. Odom, "Treatment of Behavior Disorders Incidental to Organic Cerebral Disease by Prefrontal Lobotomy," *Diseases of the Nervous System*. 13: 274-277, (1952).

[24] W.P. Wilson, "Isonicotinic Acid Hydrazide in Nontuberculous Mental Patients," *Diseases of the Nervous System*. 14: 278-281, (1953).

[25] W.P. Wilson, J.B. Parker and J. Handy, "A Symptomatic Analysis of the Effects of Perphenazine," *Diseases of the Nervous System*. 21: 340-346, (1960).

[26] W.P. Wilson, "Observations on the Effect of Toxic Doses of Atropine on the Electroencephalogram of Man," *Journal of Neuropsychiatry*. 2: 186-190, (1961).

[27] W.P. Wilson, B.S. Nashold and R.L. Green, "Studies of the Cortical and Subcortical Electrical Activity During Sleep of Patients with Syskinesias," in F.J. Gillingham and I.M.L. Donaldson, eds., *Third Symposium of Parkinsonism* (Edinburgh and London: E. & S. Livingston, Inc., 1969), pp. 160-164.

[28] W.P. Wilson and B.S. Nashold, "The Neurophysiology of Affect," *Diseases of the Nervous System*. 33: 382-386, (1972).

[29]W.P. Wilson, "Mental Health Benefits of Religious Salvation," *Diseases of the Nervous System*. 33: 382-386, (1972).

[30]W.P. Wilson and Frank L. Major, "Electroshock and the Aged Patient," *Psychopharmacology and the Aging*, eds., Carl Eisdorfer and William E. Fann, (New York: Plenum Press, 1973), pp. 239-244.

SAFEGUARDING HUMAN DIGNITY

Paul D. Feinberg
Trinity Evangelical Divinity School

The paper by professor Valenstein is both informative
and stimulating. One can not help but applaud his willingness to critically evaluate the theory and practice of his
peers. The layman finds it hard to know when claims are
realistic and when they are enthusiastic. Thus, the general public must depend upon the judgment of men like Valenstein.

Therefore, in this response attention will not be given
to the details of Valenstein's paper, but to some very fundamental ethical, philosophical and theological questions
which the procedures of chemotherapy and psychosurgery raise.

THE PROBLEM OF A MATERIALISTIC VIEW OF MAN

It is without question true that most of contemporary
philosophy and science is materialistic in its presuppositions about man and the universe.[1] There is a large body
of philosophers and scientists who hold that our so-called
"mental" states are in fact reducible to brain states. Thus,
it would seem to follow naturally that if it were only possible to locate them, then one has taken at least the first
step in controlling such behavior.

It is not the purpose of this paper to question the truth
of the theory that certain areas of the brain are connected
with various types of behavior. It is, however, my intent
to raise a flag of caution as to the results that can be expected when such materialistic presuppositions are the basis
of that hope. Both materialism and its more sophisticated
comrade, the identity thesis, are unacceptable to the Christian theologian. While theologians may and do disagree on
the exact way in which the constitution of man should be expressed, they are in agreement that man is more than simply

---

PAUL D. FEINBERG is Associate Professor of Philosophy of
Religion at Trinity Evangelical Divinity School, Deerfield,
Illinois.

his material, psychological and behavioral aspects. He is made in the image of God, and has a soul or spirit which lives after the death of his body. Moreover, unlike animals, man has both a rational mind and a will by which he can deliberate and choose in opposition to stimulation from outside.

In Valenstein's paper there seems to be evidence in support of the complexity of the human decision making process. When stimulation evokes behavior in rats, that behavior is by no means uniform. Valenstein says that when stimulated, a rat may eat, drink or show some other behavior. Similar behavior can be produced by stimulating different parts of the brain, and diverse behavior can be gotten by stimulating the same area. He indicates that "there is no correspondence between discrete brain areas and our social concerns." The aforementioned behavior has to do with rats. What could and should be said about man?

All that has been argued to this point adds up to this: If the Christian theologian is correct in his interpretation and understanding of Scripture, then the lack of uniformity in the results which Valenstein and others have gotten is not necessarily due to the need to refine existing and develop new procedures. It may well be because of the complex nature of God's creatures, particularly man. Or to put it another way, the expectations of scientists may be higher than are realistic because of their committment to materialism or the identity thesis.

THE PROBLEM OF "IN PRINCIPLE" JUSTIFICATIONS

While experimental programs are needed if science is to make progress, one must question the extent and duration to which "in principle" arguments should be used as justification for further research. These "in principle" arguments are quite common in science. They proceed along these lines: There is no "in principle" reason why a proposed goal or reduction can not be achieved. Therefore, experimentation is carried out for a period of time. Nevertheless the end or result sought is not attained. This is the case in spite of numerous attempts and varied approaches.

In such a situation it is, nonetheless, common to hear some scientists call for continued effort and money to be spent, merely because in principle there is no reason that the desired consequence can not be attained. While it is indeed important that the scientist's search should not be curtailed too soon, it does make sense at some point to question the validity of the "in principle" argument. An "in practice" demonstration should be demanded. Or, to put it another way, if wide and varied testing does not produce the desired result, is it not wise to wonder whether there is not some "in principle" reason why the goal has been frustrated?

The first two problems which have been presented deal with the issue of whether human behavior could be controlled by chemotherapy and psychosurgery. These matters have primarily been methodological. This paper now turns to two further problems that have to do with the question of whether, if it were possible, human behavior ought to be controlled. This question deals with the ethical and moral issues involved.

THE PROBLEM OF COST/BENEFIT MODELS OF DECISION

Again, from Valenstein's paper it appears that he and most scientists feel that the method whereby a decision is best reached concerning the acceptability of a procedure or a course of research is that of utility or cost/benefit. This model for decision making seeks to help one maximize goods over evils.[2] If a contemplated course of action has great potential for good, even though it may involve risks (some of which are not determinable without experiment), these scientists argue that the course of action should be pursued. At very least, it should not be prevented.[3]

Certainly there are many decisions where the cost/benefit model is appropriate and helpful, However, even when it is used in its proper place, it is susceptible to much abuse. When some pet project is at stake, tunnel vision of the type spoken of by Dr. Valenstein should not be unexpected. It is quite easy to intentionally or unintentionally miscalculate.

There is, however, a much more serious objection to be raised against this model. As an evangelical Christian, one must be committed to ethical absolutes which are universally binding. These absolutes are grounded in and a reflection of the moral character of God himself. Thus, while utility or cost/benefit analysis will be of help in some cases, the Christian must never allow any decision gotten by weighing goods and evils to prescribe a course of action which would result in breaking some ethical absolute. This is preeminently the case when man is the subject of the proposed human engineering. When one takes seriously the biblical record which states that man is the object of divine creation, made in the image of God, the recipient of God's redemptive activity in history and the possessor of an immortal soul, one must speak out against anything that would dehumanize him in life or hasten his death. This is true even though much benefit might result to society as a whole from such experimentation. Such consequences are definitely a possibility as can be seen in Valenstein's testimony that some subjects remain as vegetables after treatment.

Now, it might be argued that, in the case of chemotherapy and psychosurgery, the results are not known because these procedures are new and uncharted. There may be moral absolutes, but in this case transgressing them would not be intentional. One does not know that harm will eventuate; there is only the potential for such.

Does this put one's mind at ease? Hardly! Since man is a special creature in God's creation, the problem becomes more acute. It must be admitted that only God is omniscient. There is a certain amount of "chance" in whatever one does. But where experimentation has produced such differing results in the same animal subject, it seems only prudent and safe to assume that the potential for damage is so great that human subjects are unacceptable, whatever the benefit might be. The whole point of this discussion is not to block, even in *de facto* terms, experimentation but to limit, restrict or set guidelines for such research. More importantly, it is to caution those who would depend entirely upon a cost/benefit model of decision making, that they face serious risks in violating God's absolute commandments.

THE PROBLEM OF HUMAN AGENCY OR FREEDOM

While some of the claims like those made below by Kenneth Clark may be unjustified, they do raise again the question as to whether such results should be sought. Clark, in his 1971 Presidential address to the American Psychological Association convention, suggested that it might be possible in the near future to intervene biochemically so that the negative and primitive behavioral propensities of man might be eliminated.[4] Through the implantation of electrodes the behavior of criminals could be controlled so that they could not commit further crimes.

If such practices are at the threshold of realization, one must ask whether such "progress" ought to be implemented. As a Christian, the answer must be a resounding no. Any consequences, even the elimination of negative and evil behavior, which have as a side effect the reduction of a human being to

184

little more than a vegetable devoid of the power of choice, must be rejected out of hand. The "progress" that is the fruit of such practices is not wanted.

In this regard it should be remembered that God could have denied man the right of choice when he created him, and God can even now take away that power from man. In so doing He would have prevented the occurrence of evil. Yet, if Christian theologians are correct, God chose to allow man freedom even though that involved the possibility that that freedom might be misused.[5] If an omniscient God chose for his own purposes not to deny man this possibility, the Christian must speak out against anything that would threaten it. Does society have the right to develop biochemical and surgical techniques that will prevent the possibility of unacceptable behavior? No, they do not. To do so would be to usurp the place of God.

The burden then of this paper has been twofold. First, the problem of whether the expectations of the scientists can be realized has been raised. It has been suggested that there are at best some questionable presuppositions which make the attainment of certain results dubious. Second, it has been questioned whether the kind of chemotherapy and psychosurgery spoken of by Clark and others ought to be pursued. The answer here has been clear. Any practice or procedure which threatens the violation of the moral absolutes of God to man or would deprive man of his God-given capacity to choose is morally wrong.

FOOTNOTES

[1]R. Carnap, *Philosophical Foundations of Physics: An Introduction to the Philosophy of Science*, ed. Martin Gardiner (New York: Basic Books, 1966) and H. Feigl and G. Maxwell, eds., *Current Issues in the Philosophy of Science* (New York: Holt, Rinehart & Winston, 1961).

[2]Herbert A. Simon, *The Sciences of the Artificial* (Cambridge, Mass.: The M.I.T. Press, 1969).

[3]For a clear explanation of the problems with a cost/ benefit model see Richard B. Brandt, ed., *Social Justice* (Englewood Cliffs, N.J.: Prentice-Hall, 1962). The position that God has given men moral absolutes is ably presented in Norman L. Geisler, *Ethics: Alternatives and Issues* (Zonder-

185

van Publishing House, 1971).

[4]Kenneth B. Clark, Cited in *American Psychological Association Monitor* (October, 1971).

[5]Alvin Plantinga, "Free Will Defense," *The Philosophy of Religion*, Basil Mitchell, ed. (London: Oxford University Press, 1971), pp. 105-120.

PART VI
PSYCHOLOGICAL ENGINEERING

In this section psychologist Perry London examines
behavior control methods which influence behavior non-
coercively. In particular he focuses upon conditioning and
its implementation on a mass basis through education.
London argues that the most critical value issue in relation
to behavior control is political in nature. That is, the
treatment of individual deviance is seen as the fundamental
ethical question. He predicts that individual happiness and
choice will have to be sacrificed for the sake of future
social order. Dr. London highlights the struggle that will
be involved in arriving at a satisfactory balance between
individual volition and the employment of behavior control
procedures to bring about conformity for the sake of social
order.

Rodger Bufford agrees with London that it is in the
educational setting that ethical issues regarding behavior
control will be most centrally raised. He indicates that
the pervasive and subtle nature of conditioning techniques
make it crucial for ethical questions to be raised. In
arriving at ethical issues, Bufford points out that the
value system used is critical. Most psychologists do not
believe in absolutes; the resulting problem of such relativism is highlighted by Nazi Germany: within his culture,
Hitler was correct. The alternative that Dr. Bufford
suggests is to accept God-given absolutes, briefly summarized in the Golden Rule.

Philosopher Allen Verhey criticizes Dr. London for
failing to directly address the question of justice. He
challenges the notion that the lower limits of conditioning,
such as in child rearing, are unchallengeable. Dr. Verhey
then suggests that the issue of social power be looked at
from the standpoint of a spectrum that ranges from cooperation
to intervention to aggression (coercion). He argues that
all deviation from cooperation require justification.
Finally, Verhey argues that, although technology has contributed

187

to the democratic ethos and power of the individual, its role is now shifting. He believes that the impact of increasing specialization and technical complexity will be the erosion of democracy and volition. In its place a new controlling elite is emerging. Consequently the burden of justification for intervention must be placed in the hands of the technologists.

Psychologist Paul Clement concludes the section on behavior control with a discussion of the distinctions between process and outcome values. Process values, according to Clement, deal with how behavior is changed. Outcome values, in contrast, deal with the substance of the change; that is, what behaviors are changed and in what directions. Clement states that most of the current discussion about values and science are concerned with process values. In contrast, the Scriptures are a source of outcome or end values. Dr. Clement then goes on to argue that the application of human engineering procedures may be legitimately and effectively used to aid the Christian in reaching those end goals traditionally thought of as producible only by action of God's Spirit.

BEHAVIOR CONTROL, VALUES AND THE FUTURE

Perry London
University of Southern California

There is a bandwagon effect in human concerns, and it grows
greater as communications grow better. The technology of behavior control is getting scrutinized nowadays more due to
that bandwagon effect than because it represents something new
in human affairs. To the extent that it does involve new devices, it deserves new attention. But most of what gets looked
at with alarm or approval is part of an old array of devices
and dispositional problems that people have had since the beginning of civilization. This includes the ethical and value
problems involved, with few exceptions.

Ethical concern is currently "in," to the point that the
bandwagon virtually is giving birth to a new intellectual or
social functionary, the "ethicist." Books, papers, conferences, and even congressional committees attend increasingly
to the value implication of behavior control technology. The
problems may be old, but contemporary concern about them is
largely new.

One reason that the development of behavior control technology produces more ethical concern now than it might have
done a century ago is that people wouldn't have known about it
as much then, so there were fewer folks able to worry. Another
is that, because of the lack of instantaneous and detailed communications media, there would have been less widespread knowledge of the horrible misuses to which the technologies of supposedly humane people could be bent.

PERRY LONDON, professor of Psychology and Psychiatry, University of Southern California, has written over 100 scientific
articles and two books relating to ethics and human engineering
issues, THE MODES AND MORALS OF PSYCHOTHERAPY, and BEHAVIOR
CONTROL. Dr. London is a former Career Scientist, Development
Fellow, NIMH, and a Fellow of the American Psychological Association and the Institute of Society, Ethics, and Life Sciences, among others. He is listed in WHO'S WHO and AMERICAN
MEN OF SCIENCE. The above article was copyrighted in 1975 by
Perry London.

But the central issues of concern would really not have been much different then. A thoughtful effort to address those issues may be greatly aided by their relative timelessness and ubiquity in human affairs.

There are two most important issues, in my opinion: First, the availability of a technology has a self-stimulating impetus for us. Its uses are channelled by the value systems which guide the technicians, their mentors, and their customers. Second, therefore, the development of a technology is largely a function of the value system of a society, not merely of its physical or economic capacities.

Since behavior control technology has emerged largely as procedures for treating medical problems, most discussion of it in scientific and professional circles has concerned the technical definition of various disorders and the kinds of research, education, and practical restriction that should be put on use of the technology. Brain surgery, drugs which affect the mind, and psychotherapy have all been primary objects of concern for this reason.

But the ethical and social issues which arise directly from the proliferation of existing techniques have their main impact on mores, norms, sensibilities, and life styles outside the domain of clinical ailments. The most significant issues are moral and political rather than medical.

What we should do with the technology of behavior control is finally a question of values. We shall look briefly at some of the technologies, the value issues implicit in some uses, and what the future may be like under those conditions:

(1) The first section describes the most important behavior control technologies, often forgotten in contemporary discourse because they are the oldest, most effective, and most pervasive. These are methods which influence behavior by non-coercive, educational means, rather than by drugs or psychosurgery. The prototype of these methods, which I shall discuss here, is conditioning.

(2) The second section argues that the most crucial value in connection with behavior control is the problem of individualism or social deviance, and the extent to which social conformity can be legitimately coerced. Reliance on conditioning as the most fundamental means of behavior control implies a value system which augments volition, or the sense of personal freedom and self-control in people's lives. Some of this sense of volition must be illusory for the sake of social regulation.

190

(3)   The third section argues that such a value system
implies a utopian social order whose future dimensions, though
not entirely clear, would involve a constant tension between
individual needs and the requirements of society. Such a
social order would sacrifice individual happiness and well-
being.

BEHAVIOR CONTROL BY CONDITIONING

Most preoccupation with behavior control technology is with
drugs and psychosurgery, because they are dramatic and seem
to promise very precise means of control.

But the main themes and value problems of behavior control
are perhaps contained more in such nonphysical approaches as
psychotherapy and hypnosis, education and the communications
media, the technology of conditioning, and electronic surveillance. Of these, conditioning methods deserve particular
attention because they combine some of the precision of drugs
or surgery with the pervasiveness of education. More important is the fact that they are administered, quite literally,
with mother's milk, from infancy onward. They are the most
pervasive means by which we learn both our emotional attitudes
and our skills. Finally, they are promising paradigms for the
resolution of some of the value problems of behavior control.

The term conditioning is sometimes used synonymously with
behavior control. One reason for this is that, with conditioning methods, nothing is ostensibly controlled but the behavior itself, rather than the chemical or physical body
processes which underlie it. This notion is, of course inaccu-
rate - the physical substrates of behavior are real. But the
illusion that they are psychological without also being physiological deludes some people into thinking that they are not
as potent.

Conditioning is the most important behavior control method
because:

(1)   it is the most pervasive and probably always will be;
(2)   in some respects, it is the most irreversible;
(3)   it is the least obviously coercive;

This is not an appropriate place to discuss the details of
conditioning technology, but it may be worthwhile to review
briefly the processes involved and the aspects of behavior to
which they apply. Broadly speaking, there are two kinds of
conditioning process, classical or Pavlovian conditioning and
instrumental or operant conditioning. Both were discovered at

the turn of the century, in Russia and America respectively.
Both have been subjects of endless study since. In evolutionary terms, classical conditioning may be considered a
more primitive process than instrumental conditioning because
it is found in lower species than the latter. Both are of
utmost importance in human learning and behavior.

In general, classical conditioning may be viewed as the
training of sensory processes by which a person learns to
connect a familiar response to a new stimulus. Control is
established over internal, involuntary behavior like emotion,
mood, sensation, and the functioning of glands and of smooth
muscles in the stomach, blood vessels, and heart. Instrumental conditioning may be viewed as the training of motor
processes, by which a person learns to make new responses to
familiar stimuli. Control is established over voluntary behavior, including social, intellectual, and psychomotor skills.

In combination with each other, and increasingly with drugs,
surgery, computers, and electronic gadgetry, the two conditioning methods may be used to teach control over virtually
any attitude, any skill, or any emotional tendency. Love,
hate, psychosomatic responses, anxiety, anger, and the emotional component of attitudes and meanings, are all the products of classical conditioning. Problem solving behavior,
the way we puzzle things out, study, and develop our skills
and action tendencies, ranging from learning to walk to learning skiing, tennis, dancing, and fighting all evolve from instrumental or operant conditioning.

The modern technological aspects of conditioning that make
it more powerful than in the past, are the things which make
its application systematic. These include the discovery of
antecedent conditions, internal (i.e., motivation) and external
(i.e., distractions, attention getters), which implant it most
effectively, and of the reinforcers which take most quickly
and last longest. Drugs and computers are the two biggest
contributors to these respective dimensions. Drugs change
aspects of motivation, in the broadest sense, such as arousal,
and computers improve feedback.

The greatest potential of these technologies for behavior
control lies in the rearing of children, which is historically
where they have been casually applied. Today, incorporated
into behavior modification, they are used for teaching troubled children verbal and motor skills, control of tantrums and
bedwetting, alleviation of some of the disabilities of mental
retardation and, hopefully, the treatment of autism. Their
implications for broader use in the rearing of children extend

to the concept of Utopian society developed in WALDEN TWO
(Skinner, 1948). There, every aspect of life is carefully
engineered with operant methods. Self-control is the most
important ethical virtue, and the community's planners train
the children in self-control through a graded series of frustrating situations. In school children are not "taught" any
subjects. Instead, they are given a survey "of the methods
and techniques of thinking, taken from logic, statistics,
and scientific method, psychology and mathematics....They
get the rest by themselves in our libraries and laborato-
ries."[2] No deliberate motivation for diplomas, parental approval, or high grades are built into the children. They are
motivated simply by "natural" curiosity.

Notice that WALDEN TWO has exactly the opposite of the
child rearing practices Jeanne Jacques Rousseau thought of
as producing the "natural" man out of early childhood. While
he argued that children should be essentially left alone,
Skinner proposed that they be systematically controlled in
order to teach them systematic self-control. Once it is
learned they can then do pretty much what they please.

Skinner's conception, published a generation ago, does not
take account of the uses of classical conditioning in combination with instrumental conditioning, nor their combination
with computers, drugs, psychosurgery and electronics. But
the principle he establishes in WALDEN TWO, and has repeated
in all his most important publications, is still the same.
That is, you cannot have a satisfactory social organization
unless you make optimum use of the possibilities of behavior
control technology. The point must recur repeatedly because
it is the heart of the issue of applying whatever values mo-
tivate our uses of behavioral technology in the first place.

Behavior control through conditioning has had varied applications to adult behavior disorders. The two areas where
it has received most application are the very ones that put
in glaring limelight the value problems that must be faced.
These are mental illness and crime, the two major arenas of
social deviance which elicit continuous tension between individuals and society. Three applications of conditioning have
been most important here: first, the establishment of behavior modification and related rehabilitation programs in prison
settings; second, the use of token economies within mental
hospitals; third, the treatment of sexual problems, particularly homosexuality.

In all three cases, the problems of value which have arisen
have gone far beyond the issue of the efficacy of the methods.[3]

Indeed, the methods have been only moderately effective, at best. The question of whether they should be used, under what circumstances, by whom, on whom, and how, does not assume that they are very precise or very potent, but only that they are available. We shall return to this matter because it clarifies that the value issues of behavior control which require external legislation and regulation are those in which the technology involved enables some people to exercise power over others, whether or not for the benefit of the victims.

Briefly, token economies (and most prison rehabilitation systems) are closed, total institutions operated like businesses. Inmates are rewarded for desirable behavior by tokens, which can be used as money, or by privileges which make life more pleasant for them within the institution. People's behavior is shaped this way partly because they learn to repeat acts which earn rewards, no matter what the rewards are. More important, tokens become valuable by the fact that necessities as well as luxuries must be paid for by them. In a token economy hospital ward, patients may have to pay to eat, to go to bed, to shower, to sleep, even to use the toilet. In a prison, good behavior earns comparable material rewards, and gradually increases the relative freedom of the prisoner to the point where leave or release eventually occurs.

The treatment of sex problems includes a variety of conditioning methods, but the treatment of homosexuality in particular, by means of aversive conditioning in particular, has been scrutinized more carefully and caused more disapproval than any other kind. In general, this is a combination of providing painful electric shock or other painful stimulation in connection with homosexually provocative stimulation, and removing such painful occurrences in face of heterosexual stimulation. The results of these treatments have been mixed. Success has probably been more dependent on the motivation of the patient to get rid of homosexuality than on the potency of the shock-reward regimen.

There has not, as yet, been very extensive effort made to apply conditioning methods to a great variety of other adult problems. However, biofeedback technology, a variant, has promise for the treatment of some psychosomatic and physical ailments, such as migraine headaches and high blood pressure.

The big value problems which arise in relation to these methods do not apply so much to the clearly medical conditions indicated here. Rather, they arise when using conditioning for the kinds of social deviance to which it is already, however crudely, being applied right now. Suppose, for instance,

194

it were possible to systematically manipulate the attitudes of criminals towards obedience to authority, or to produce massive anxiety or nausea, a la A CLOCKWORK ORANGE, to the urge towards violent or criminal behavior in oneself. Or suppose one could dependably improve psychotics, or mental retardates, to the point where they could then assume minimal functioning outside hospital walls. They could then be sent back to indecent, and intolerable family or social surroundings, so that they could be effectively enslaved and abused without also being a public burden. Or suppose the treatment for homosexuality were absolutely successful and equally applicable to heterosexuality, so that people who sought the peace of celibacy could be guaranteed to find it.

As with all the behavior control methods at issue, the values which promote the maximum use of conditioning technologies, and the values which the maximum use of those technologies in turn promotes, are those of reducing individual pain and of enhancing the sense of self-control, that is, of personal freedom, in people's lives. That sense comes, however, from being satisfied with one's behavior, not from being capable of altering it.

One cannot challenge the lower limits of the use of conditioning (nor, indeed, of drugs or surgery either) unless one also wishes to challenge the lower limits of education and socialization. All three are the same, at least in infancy and early childhood. While there may be good reason to debate whether we should socialize our children as we do in some respects, most of us are so committed to the virtues of toilet training and language development that we would not be willing to forego them at any cost.

For everyone, however, the value question rises with respect to the upper limits of the use of these things. This is especially so in situations where it is possible to use other means of control, or no control at all, without destroying society.

The tensions of importance occur in two situations: Where the individual wants things for himself which are at odds with society's norms, and where the society wants things of him to contribute to the general welfare. To see these clearly, we will look at the theoretical nature of politically permissible social deviance and at the practical matters of crime and mental illness.

BEHAVIOR CONTROL AND SOCIAL DEVIANCE: THE VALUE ISSUE

The attitude of an organized society toward the uses of

195

behavior control is a special case of its general moral outlook on the relation of the individual to the group. Societies
are naturally conservative; their functions cannot be fulfilled unless people can be depended on to adopt common public
practices and to avoid deviating from them unexpectedly. For
the group establishment to serve the interests of its individual members, they must share enough common cause to permit
an establishment to work.[4] Social organization, it seems,
is purchased only at the cost of the ease with which individuals could once dissent or withdraw from the group. Eventually, customs become laws, plausible conveniences turn into
moral proprieties, and trivial habits into public manners.
This holds true until any irregular and deviant public behavior by individuals may seem to threaten society and create
suspicion and alarm.

A society's potential sufferance of individual dissent
depends partly on its population density and technological
development. A small technically primitive social organization cannot function very well if any of its members fail to
do their assigned jobs, regardless of the reasons for their
failure. If the food-getting member of a three-man society
goes on strike, for example, the consequences for the others
may be disastrous. Their demands that he resume work will
not be stilled by his insistence on his individual moral prerogatives. In a technically advanced society like ours, however, a given person's refusal to join the army, pay income
taxes, or even work for a living, has so little effect on
the conduct of the nation's business that any decision to in-
dulge or punish him can rest more on the abstract rights and
wrongs of the case than on the assessment of any material
damage he inflicts on society. In general, a primitive society must treat individual deviance as a functional problem
while an advanced one can treat it as a moral problem.

The economic and political history of the United States
make it inevitable for social deviance in general, and mental
illness in particular, to be moral rather than functional
problems in this country, despite any damage they do to our
gross national product.[5] A political ideology of individualism that officially authorized the pursuit of happiness,
exonerates the entrepreneurial spirit and, by implication,
invites people to be as free of society as they wish, is
ideally suited to social deviants. The viability of such a
society depends on its wealth, on there being few enough deviants so that they can be supported, on a relative dearth
of external enemies, and on a lot of good luck. The United
States has had all this in abundance. The favorable

combination of technology and ideology has brought us to a position where we can afford deviance economically. However, we may be endangered morally by the very freedom from social responsibility that is implicit in our political ideas.

The political ethics of individual liberty in America evolved in times and under conditions in which the deviant public rights which people sought were in fact minimal. Many potentially deviant groups, moreover, were too weak to forcibly demand such rights for themselves. The quest for religious liberty, for example, always involved churches and practices which were not unthinkably different from those of the majority. No witches' coven sent representatives to the Constitutional Convention to assure a separate Church and State. The demand for representative taxation involved no radical departure from prevailing views of the nature of property or the proper distribution of wealth. No communists demanded the nationalization of land at the Continental Congress. Jews were not only not very deviant beside Protestants, they were also not very populous. Negroes were not yet officially people. Witches, vegetarians, nudists, and homosexuals, let alone pedophiles, necrophiliacs, and women's liberators, were not notably vocal.

In short, the early politics of individual freedom in the West were based on an approximate consensus of a fairly homogeneous population. There was no fear that there might someday be too many people wanting too many different kinds of freedom of too shocking a kind and in too much conflict with each other. It was to be around the extent and quality of deviance and of conflict that a political issue would be joined.

Such an issue has been joined, I believe, in connection with behavior control. The technology of behavior control makes it possible today to exact individual conformity with greater reliability and less risk of resistance than ever before. Such a threat to traditional ideas of personal responsibility and political liberty makes it necessary to reconsider many of the specific laws and general mores which our society still uses or endorses to limit individual conduct.

Historically, the coercion of individuals has been limited by the machinery for implementing coercion, which was crude and undependable. Today's control methods, however, increasingly make possible the effective engineering of consent. Today we can envision a time when people can be dependably moved to act responsibly (that is, the way the controller

197

wants them to) and to be happy in their own compliance.
These developments do not change any fundamental issues in
the control of human behavior, but they sharpen them.

To be able to eliminate personal license and still leave
people with the feeling that they are free may sound ideal
for the purposes of correction and rehabilitation. But it
does tend to unhinge the conventional political morality on
which modern democracy is based. Democratic political theory
regards the coercion of behavior or consent as the essence
of tyranny. But our common notion of coercion presupposes
that, in losing freedom of action, people do not also "lose
their heads", especially part of their minds which carries
the consciousness that they have lost freedom of action. You
can make someone say he wants to obey you, of course, by putting a gun to his head. But you cannot make him mean it.
When the gun is gone, the illusion of consent goes with it.
The victim of coercion will then say, and do, what he pleases.
But this reversion to type is exactly what modern control
technology promises can be eliminated.

The issue is not novel, but it must arise more and more
frequently today in relation to certain concrete behavior
problems because of certain new conditions of life. First,
societies have never been so big, so densely populated, and
so free of the burdens of productive labor and defense. Second,
the opportunities for social pluralism and for individual deviance have never been so many. Third, the potential for both
rehabilitation and repression has never been so great. (These
terms may be synonymous with respect to action, and differ
only in perspective or intent.)

What is the "common welfare" or "common good"? How desirable, useful, and valid are these concepts? Why should
they compel the loyalties of individuals indeed obligated to
participate in society? How, and how much must they pay for
them? Most critically, what penalty may the commonwealth
exact for failure to comply? Are there legitimately applicable standards of public conduct, infringement of which is
legitimately subject to punishment? Finally, what bearing
does the public claim on the individual, and the availability
of a technology to exercise that claim, have upon the fulfillment of the individual's life goals? What should those
goals be? Can they be prescribed for him, or can he be pro-
hibited from seeking his own?

I submit three propositions in this connection. Taken together, they represent a volitional thesis for using behavior
control technology as a means of social control:

First, in our society as it is, the political preconditions of individual freedom can no longer rationally include a compensating positive obligation of the individual toward society (as opposed to refraining from materially damaging it). In effect, social deviance is more viable now, and thus more defensible, than at any previous time in human history. Thus there is less excuse today than ever before in history for demanding that someone conform to social custom because the welfare of society supposedly depends upon it.

Second, most of the ethical and legal problems of behavior control technology revolve around practical definitions of crime and mental illness in our society. The preservation of individual liberty requires that we reconsider these categories. Perhaps we will need to strike from them some of the individual acts we now treat as crimes or mental illness. If we stopped defining certain acts as crimes or mental illness in the first place, we would not even think of forcing people to control them.

Third, a liberation ethic may sometimes prove less productive of individual happiness than would an ethic of benevolent coerciveness. This is possible even though it supports a fuller variety of human activity, with both the risks and riches which are entailed. Changing in a more liberal direction may be necessary for the preservation of personal liberty, but it may only be trading off one misery for another in the lives of some people who are made more free.

THE CRITERION PROBLEM

Identifying the value issues of behavior control in terms of the social problem involved does not alter any concern with the issues of individual morality or theological interest. These issues underlie the questions of how people ought to behave in the first place. But the central fact of behavior control is that it involves somebody doing something to somebody else. For these purposes, it is a question of human or social engineering. The see-saw ends of the rights and wrongs of all such interactions are coercion and volition, nothing else. The criterion problems involved are those of who shall do what to whom under what circumstances. The modern prototype of doers is mental health experts, of those done to the mentally ill or socially deviant. The circumstances defining the proper conditions have been elusive. Together, these constitute the criterion problems of behavior control.

There are some elementary and fairly good criteria for most

199

conditions that most people call illness. For one thing, the person who has it usually hurts. For another, it is generally difficult for him to work or otherwise function when he is ill. For a third, he is likely either to recover from it or get worse in reasonably short order. If he recovers, either there will be no trace of his having ever been ill or he will retain some visible stigma, defect, or disability that shows rather clearly that he was sick. If he does not recover, he will probably die of the malady, again clearly suggesting that he had been ill.

In the case of mental illness (from the professional rather than the social view), nothing could be less clear more often than what defines the condition.

At the present time, it is generally conceded that, whatever anatomical or physiological basis may exist for psychological disorders, the disorders themselves are mostly reflections of troublesome learning patterns. These are usually acquired in childhood, and tend to be sustained by the exigencies of the person's contemporary life. But this learning approach to disorders leaves the healer, now perhaps educator, with the question: If the troubled individual is not "ill" but has only learned some behavior pattern that is unacceptable to most other people, if not to him, who is to be responsible for dealing with it? If the patient does not think he is sick, then why should the doctor be obliged to make him well? It is true that doctors often function as public health officers and authorize quarantine and other restrictive procedures. However, they do so only where there is clear and present danger to everyone if the sick person walks around untreated. Contagion is not a common problem with psychiatric disabilities.

THE MORALS OF SOCIAL ENGINEERING

The mental health expert faces the criterion problem most dramatically where the patient is not hurting because of his condition but some social convention. A related difficulty occurs when the patient is suffering more from the social stigma attached to his condition than from the condition itself. Homosexuality is a good example. Many homosexuals are less disturbed by their sexual behavior than by the social scorn they feel because of it. In such cases, discretion about what treatment should accomplish often rests with the psychotherapist. He may find it easier to free the homosexual of anxiety about social disapproval than to change his sexual style. Such a course is easy to justify if the therapist himself does not truly regard homosexuality as

200

immoral or injurious. At the same time, however, the operations he then attempts make him a social engineer. It may give him legitimate cause to think that he is thereby deliberately working to reconstruct the sexual mores of society.

The potential of mental health experts for social engineering is quite large, although relatively few of them are publicly concerned about it. Homosexuality itself is a statistically common problem for American men. Other problems of sexual conduct and morals have also been increasingly confronting psychotherapists in their routine dealing with patients. With the exception of those therapeutic personnel whose religious or institutional affiliations make it easy to recognize the conservative bias of their outlook, it is clear that most practitioners in most private offices and outpatient clinics are liberal in their own moral judgments of sexual freedom, if not in their own private activities. At the least, this means they support a changing sexual morality that is permissive of extensive sexual activity prior to marriage. They frown on sexual deviancy mostly because of its social stigma alone and would not have it outlawed. They even judge most norms of sexual restriction with concern chiefly for the psychological rather than other consequences of their breaching. At least one formal study has shown that psychotherapists tend to fall ideologically among the most Bohemian groups in American society.

Most practitioners would probably say that the positions they take on any kind of problem depend on the unique attributes of the situation and the patient being served. Their work is well done if they are able to change the person's life in some beneficial way. This usually means in a way that makes it more pleasant. The fact that many people suffer in the same (non-unique) way and can be changed by the same (non-unique) treatment does not usually make the therapist feel like a social engineer. As long as he has not gone into collusion with his colleagues to produce the changes for which he is individually responsible he can maintain this view. But it is hard for him to see himself as anything else when, in concert with them, he deliberately changes the psychiatric status of the conditions he deals with, as the American Psychiatric Association recently did with homosexuality.

In so doing, we see how circumstances intervened in the technical encounter of doctors and patients, of controllers and those controlled. For it is the temper of the times which enables homosexuals to press the case for their own mental health. Doing so with respect to health alters willy-nilly the moral climate that motivates homosexuals to get changed.

in the first place. At the same time, it rearranges the agency of the therapist so that he now further detoxifies the former social stigma of homosexuality. He now makes its alteration a hygenic, even a cosmetic-elective, "condition" rather than an ailment.

It is no longer a socially radical bias of psychiatry which views self-fulfillment as the goal of treatment. (This may sound like the same thing as adjustment, but it is far from it. Adjustment implies that the object of treatment is the comfort and well-being of the patient, whether conservatively achieved by conformity to society, or liberally left for him to decide. The idea of fulfillment, however, means that the patient needs to become either a different kind of person than he is or a lot more of it. That is, he needs to achieve the maximum personal maturity of which he is capable.)

The important point is that the whole chain of circumstance and interaction, tied together as it may be by a Gordion knot, is initiated by the value attached to homosexuality, not by the technology for dealing with it. The technology would not have been invented in the first place were the condition socially accepted. The technology's use will now be curbed because the value attached to the behavior has shifted, in society and psychiatry, not because the technology is good, bad, or indifferent. So the future, like the present, depends more on the values of our society than on the ingenuity or intentions of its technologists.

THE FUTURE OF BEHAVIOR CONTROL

The problems of the future with respect to behavior control are those of repression and liberty. Both are problems of utopia. Repression speaks for itself; a repressive social system will do what its rulers want and will serve their purposes. But a gentle utopia will have the problem of balancing individual desires and liberties against social needs in situations where the agents of society have the power to satisfy either one. To the extent that they place a high value on pain avoidance, they will opt for control methods which maximize the individual sense of volition, control, incentive and reward. In effect, then, they will promote conditioning in general and behavior shaping in particular as maximally desirable.

It is only interesting to make predictions about the future of behavior technology if American society develops in a nonrepressive political direction. If political repression becomes widespread, it will be used in whatever ways will contribute most to the maintenance of a tyrannical government. In a nonrepressive America, however, the general effect of behavioral technology will be to increase people's feelings of having choices about how to live. Public concern about the misuses and dangers of control technology is largely a function of unfamiliarity. It will naturally diminish in a social context where its applications are made at the discretion of the people it is used on, rather than at the pleasure of coercive agencies of government. In such a situation, one can foresee the following results with drugs, psychotherapy, and psychosurgery[6]:

Drugs will be removed from prohibitive control and will either be legalized, and then medically controlled, or not legalized and not controlled at all. That is, they will be controlled by the demands of the marketplace. In any case, they will be removed from police jurisdiction. The social issue of drugs is mainly over drugs that you take for reasons other than being sick. This includes familiar things like psychedelics (Cannabis, LSD, peyote, etc.) and energizers, but also drugs that should be available fairly soon for boosting memory, effective aphrodisia, and cheap, certain and safe means of committing suicide.

Psychotherapy will become more popular, and will divide more sharply into technical branches like psychoanalysis and behavior therapy, on the one hand, and into quasi-religious, quasi-recreational activities like encounter groups, on the other.[7,8]

Psychosurgery in general, and ESB in particular, will be relegated to the technical specialities they ought to be. As Herbert Vaughan has reported: "Many of the problems of ESB are uniquely medical." The potentialities of this method present more drama than social danger. This means there will ultimately be more coercive use of them by government on a very restricted population of social deviants that cannot be as well controlled by other means. This will come about by common consent. Often it will be at the request of the people to whom it is done, especially in cases of uncontrollable sexual and aggressive impulses. Its use will be popularly viewed as a kindness. It may mean the end of jails for some things. In most situations, ESB will be a form of socially elective surgery in which the internal professional controls over decisions to use it will be more restrictive than at present. To the extent that these methods can be applied to the improvement of IQ, memory, or sex drive, it is possible that there will be some requests for it by patients themselves, as is the case with much cosmetic surgery today. If drugs offer such possibilities for self-imporvement, however, as seems likely, this will not happen to psychosurgery. The public does not like getting cut.

The indirect social effects of behavior control technology are part of the general impact of technology on behavior. Increased leisure and broader education have given people expanded horizons and shrunken restrictions on their own lives and actions. People know more than they used to about what is going on in the world and what they can do. They try more different things, and they "fool around" more than ever before in history. Contraceptive pills are the best example of such effects; they have little direct effect on sexual appetites, but tremendous effects on the motives and opportunities for sexual behavior. The same could be said of the miniskirt, the automobile, and Dr. Kinsey's prodigious scholarship.

The social consequences of psychotherapy and drugs will come from the waves they make in multitudes of individual lives and in general fostering of self-preoccupation as the source of value judgments. They will thereby shift individual morals and mores increasingly in the direction of hedonistic goals. Barring its political misuse, the social consequences of psychosurgery will be much more abstract. Its use will stimulate intellectual conversation about the nature of man, of will, and of freedom. Some of the moral traditions of our society will be undermined by its fostering of skepticism about the assumptions on which they are based.

Psychotherapy contributes to preoccupation with self, the existential orientation, in three main ways: First, by its secularism and scientific pretentions, which attract people who are guilty, distrustful, or contemptuous of the values of their parents, priests, or doctors. Second, by its methods of enhancing self-awareness, which are easily transferred to relationships outside the consulting room. Third, by the self-preoccupation of therapy itself, which translates easily into an ideology of self-realization or of commitment which demands no outside loyalties. This shifts the standard of self-evaluation from that of an external morality to one of personal integrity.

The upshot of the therapeutic encounter is the recommendation that one should "do his thing", a message already pressed on him by other liberalizing forces in our society.

In the social ambience of freedom in interpersonal conduct which results, therapeutic games, the social byproduct of psychotherapy, may expand enormously. They will be expressed in encounter groups of increasingly outright unconventional objectives - nudies, feelies, and the like, rationalized and justified in arguments ranging from the quasi-medical and quasi-religious to the quasi-orgiastic. The function of all this will be to provide "instant intimacy." It will foster rapid social intimacies in a society too mobile for neighbors and relatives to fill that role. The orientation of all these groups will be increasingly towards self or peer group objectives.

As for their indirect effects, drugs fall into the general technological category of self-controlled, high precision control devices. In this respect, they are more like the automobile than like psychotherapy. Just as it would have been hard to see, in 1905, that the automobile would have, among its dramatic indirect effects, the creation of the ecology industry, it is hard to forecast what the indirect effects may be of such hedonic drugs as aphrodisiacs and memory boosters. As indicated, however, it is clear how powerful indirect effects have come from contraceptive pills, the prototype of this category: They vastly altered sexual mores because, for the first time in history, they put the control of pregnancy entirely in the hands of women. This in turn promotes the alteration of marriage customs by reducing the economic dependency of women and, therefore, lowering their incentive to remain bound up in unhappy marriages. The ease of divorce and remarriage increases in turn. One of the main consequences is, that a majority of American children are now being raised by more or less than two parents, one or more of whom did not sire them. The ultimate effect on the nuclear family, and on society, is not clear yet, but it cannot be small. While "the pill" is not, by itself, responsible for all of this, it is not irrelevant to it either. Rather, it is a significant contributor to the impact of technology on behavior.

The indirect effect of psychotherapy, in short, are to encourage self-reference as the final standard of judging one's behavior; drugs provide one means for indulging that standard. Therapy says, in effect, do your thing. Drugs give you some new things to do.

All of the foregoing has been extrapolated here as a kind of "line of least resistance" for the future of these behavior control devices in a free society. But the freedom of society is the main point at issue in people's concern with behavior control. Nothing could make this clearer than examining the long range meaning of behavior control technology.

THE LONG RANGE ISSUE OF BEHAVIOR CONTROL

The essence of modern behavior control technology, and the

ching that makes it problematic, is that it promises the ability to plan precise intervention in people's lives. Were there no promise of precision, there would be less alarm. When we talk about the social issues and implications of precise behavior control, we are talking about the prospects of utopian society. It would be useful to think about it in those terms, because that is where the thrust of this whole technology goes. All of us want it who want any serious degree of economic planning, social welfare, or the containment of random violence.

But in this connection, we often find ourselves talking out of both sides of our mouths. We speak of freedoms, including, as Gerald Klerman has put it, "freedom for the pursuit of happiness," but there is not utopian society which does not restrict freedom. By its very nature social planning may require the planning of incursions on freedom of individual conduct. Freedom means that a person has power over his self. Behavior control means power over others.

The more there is a blanket program for utopia, the more certain it is that incursions will be made. No single sane principle emerges for balancing this situation in terms of what we now cherish as individual liberty. The best approximations are principles such as reward or positive reinforcement, nonviolence and noncoercion. These are all restatements, one way or another, of the classical golden rule.

Even then, the society which may emerge will be full of constraints on the individual. The best description I have seen of such a society is given by Ira Levin in a recent novel entitled THIS PERFECT DAY (1970). In it, he describes a world united in peace and coordinated by computer. The "family of man" is a harmonious interdependency, where the moral tonus of the society is based on love, and "hate," "fighting," and the like are curse words. If anyone deviates from the norms of love and mutual consideration and commitment to the common welfare, he is given treatment, not punishment. And the treatment is genuine, a tranquilizing injection which makes him feel good and happy. Even here, of course, there are dissidents, and they persuade the novel's protagonist to join them in opposing the system.

The novel's protagonist finally succeeds in bringing down the gently repressive Utopia which reared him. But the novel ends there, for novelist Levin does not have any special social or political order to replace it. Nor do any of us, I think, who value freedom, on the one hand, and who cannot

206

honestly say, on the other, that it is evil for a society to do away with crime and overpopulation and starvation and violence and war, and to thrive on norms of love and mutual cooperation.

The question is one of how to align those norms with volition, how to bring them about as products of desire, incentive, positive reinforcement and the sense of self-control, rather than as by-products of coercion or seduction.

All of these questions are ultimately questions of means, not of ends. If people can be enjoined, by means which leave them the sense of option, to do the things that we require of civilized human beings (literally and politically), then we must be satisfied. There are no absolute ends in modern, complicated, technological, or democratic society. And all means are absolute, once applied.

The long range value issue of behavior control is the issue of the distribution and diffusion of power in society. Power can be wielded in a kindly way. Power can be cruelly used. The question is how much power over their own behavior must the good society vest in each of its members, and at what risk to its own integrity? None of us, if we are honest with ourselves, will forego the vision or the search for that society.

FOOTNOTES

[1]P. London, *Beginning Psychology* (Homewood, Ill.: Irwin-Dorsey Press, 1975a).

[2]B.F. Skinner, *Walden Two* (N.Y.: Macmillan, 1948).

[3]P. London, "Behavior Modification," In J. Schoolar, (ed.) *Research and the Psychiatric Patient* (N.Y.: Brunner-Maisel, 1975b).

[4]P. London, "Social Issues of Behavior Control: Present and Prospective," *Lex et Scientia*, (1972b), pp. 9, 132-138.

[5]P. London, *Behavior Control* (N.Y.: Harper & Row, 1969).

[6]P. London, "Legislating the Brain: The Citizen as Patient," *The Columbia Forum*, (1972), 1, pp. 2-7.

[7] P. London, "The Future of Psychotherapy," *Hastings Center Report*, (1973), p. 3.

[8] P. London, "The Psychotherapy Boom," *Psychology Today*, (1974), pp. 8, 33-38.

ETHICS FOR THE MASS APPLICATION OF BEHAVIOR CONTROL

Rodger K. Bufford
Psychological Studies Institute

In his address on behavior control Dr. London suggests
that it is in education that the ethical issues regarding
the control of behavior will be most crucially joined. He
proceeds to suggest "the prototype of (educational) means
is conditioning." In my opinion London is right, and for
several reasons. First, the techniques of conditioning and
education are already with us, while much that is discussed
in the areas of genetic control and electrical brain control
involve possible future developments. Second, the educational and conditioning procedures are already practiced on
a widespread scale. Third, it is these techniques which are
likely to be employed to persuade or influence people to
accept and use developing techniques of chemical, electrical
and genetic control.

One person aptly described one side of the issue when he
suggested that we need not fear being controlled by electrodes implanted in our brains; what should concern us is
the man with the gun to our head forcing us to submit to the
installation of the electrodes in the first place. He did
not mention it, but we should also be concerned about the
subtle influence of various educational and conditioning
techniques. These may lead us to adopt the practice of implanting electrodes, taking drugs, and so on, for ostensibly
"good" purposes, thus facilitating their exploitation for

RODGER K. BUFFORD is Associate Professor of Psychology
at the Psychological Studies Institute, Inc., Atlanta, Geor-
gia. His specialization is in clinical/experimental-child
psychology. Dr. Bufford has published articles on imitation,
self-paced instruction, and psychopathology. He has served
as reviewing editor for a number of journals and books. Dr.
Bufford is a licensed psychologist in the state of Virginia,
and is listed in current issues of PERSONALITIES OF THE
SOUTH and WHO'S WHO IN THE SOUTH AND SOUTHEAST. He has previously taught at American University and at Huntington College, where he was Chairman of the Psychology Department.

malevolent purposes. It is the more general application
of these educational and conditioning procedures which will
be our concern here.

It is perhaps fitting that the occasion for this conference coincided with the fiftieth anniversary of the Scopes
trial. The basic issue of the trial was whether a majority,
in our pluralistic society, could use the educational system to control access to information of minorities who do
not accept their view. Specifically, the Scopes trial
dealt with the question of whether or not evolution could
be taught in the Tennessee public schools as an alternative
to creationism in accounting for the origin of man. In the
intervening fifty years there has been a dramatic shift. The
issues which are now stirring controversy in California
and elsewhere are over whether creationism can be taught
as an alternative to evolution. Unlike the issues related
to electrical brain control, these and related issues are
with us now. On the one hand, this means that we have
often established casual means of dealing with them and do
not fear the techniques. On the other hand, it implies
that we are probably less watchful for their potential abuse.

The danger in educational and conditioning techniques
lies not only in their ubiquity, specificity and effectiveness, but also in the fact that they can be used surrepti-
tiously. For example, a husband might set out to alter his
wife's behavior in a direction she deems undesirable in
such a subtle manner that for a long time she remains oblivious to the process. By the time she becomes aware of what
is happening it may be rather late to institute countercon-
trol measures. Similarly, the school system may be used to
subtly teach world views and value systems which are not
consistent with those of the community.

It is this pervasiveness and subtlety of behavior control and conditioning techniques which make the ethics of
their application so critical.

The issues which arise include: who shall influence
whom; by what techniques; to what ends; what roles does the
person whose behavior is altered play in the decision process; how do we resolve the dilemma of apparent conflicts
between individual rights and social welfare; are there any
absolutes? As London has accurately noted, these issues do
not become salient in the doctor-patient type interactions
in which conditioning techniques were first systematically
and self-consciously applied. In such a context the doctor
serves as the patient's agent in a more-or-less contractual

210

arrangement by which both benefit. The issues become crucial whenever there is the opportunity for exploitative
application of behavior control techniques for the benefit
of the controller and the detriment of the controlled. In
another context Ferster has termed this paradigm "arbitrary
reinforcement" (Ferster, 1969).

ABSOLUTES

First, let's address the question of absolutes. Most
behavioral psychologists who have addressed this issue seem
to conclude that there are no absolutes. Skinner assumes,
without justification, preservation of the species as the
summum bonum (Rogers and Skinner, 1956). He goes on to suggest that we can assess empirically what cultural practices
contribute positively and which are detrimental to this
objective. In the first place, Skinner has copped out when
he fails to justify survival as a goal. Moreover, science
cannot help us decide whether adopting certain practices is
good or bad. This is a value decision which lies outside
the realm of science. Science frequently does an admirable
job of describing the immediate effects of certain interventions; but some of the consequences are often neither
immediate nor readily apparent. History is replete with
examples which suggest that what appears good at the time
may on later analysis prove to be a mistake - witness the
belatedly discovered adverse effects of DDT.

Dr. London argues for a value system that "is one of
maximizing choice, i.e. the sense of personal freedom and
of self-control in peoples's lives." He notes that some of
this sense of freedom and self-control must be illusory in
order to serve the needs of the society. As with Skinner's
value of survival, there is no justification given for this
value by London, nor is it clear how it will be accomplished.

Most psychologists who have dealt with the issues of
ethics and values fall into two camps. One includes Skinner and London. These people argue for some concept of the
highest good which is universally applicable but they are
unable to present any convincing rationale for others to
subscribe to it. The other camp consists of those such as
Ullmann who argues that the standards of the culture in
which one resides are the basis for ethical and value decisions. One is responsible to his society for his actions.
By implication, ethics and morality vary with time and place
according to these individuals. The view that the standards
of the community determine what is right or wrong is known
as cultural relativism. This view implies that for his

culture, in his time, Hitler was right.

There is a third alternative: belief in God-given absolutes. Most succinctly put these absolutes are expressed in the golden rule, "do unto others as you would have them do unto you." In research and application of behavior control techniques with those who are unable to make truly informed consent this rule must be our guiding precept. In putting it into application we must keep in mind that while we cannot be held accountable by that retarded child, we must account to society for what we do, and ultimately to the God who made him and us.

Dealing with competent adults along the traditional lines of informed consent usually presents relatively little difficulty. In the unusual case where two consent together to establish behavior which departs from society's standards they are ultimately both held accountable. The problem arises when the individual cannot truly be informed or legitimately given consent, and secondarily, when that consent is coerced. These are the cases of the retarded, the psychotic, the prisoner, and universally, the child. While most of the heat regarding ethical issues of behavior control has been generated in regard to treatment of the criminal and the "mentally ill," the issues of behavior control may ultimately become most critically focused on the area of education and child rearing. Some, such as Roger McIntire of the University of Maryland, have already seriously proposed licensing of parenthood, and suggested that training in the application of behavior control techniques be a prerequisite to such licensing.

The other side of the picture is that the public schools are increasingly reaching out further from the "three Rs" into areas such as health education, socialization and sex education. These have traditionally been the parent's responsibility. How do we protect the rights of parents and children who do not accept the same values and standards as those of the school system? The issues most salient here are the ends toward which behavior control techniques are applied, and who makes the decision regarding those ends.

The viewpoint of the conference commission is that the ends toward which behavior may be altered must be guided by absolutes of God's revelation in Scripture. What is needed is for those who hold such a view within the scientific community to enunciate the implications of these principles within their own disciplines and communicate these ethical guidelines to their colleagues and the public. It is

212

unfortunate that in the past the professional evangelical
community has largely been silent on these matters.

MISCELLANEOUS ISSUES

With regard to the issue of conflict between individual
freedoms and social welfare, it seems to me that, properly
understood, the welfare of society and the individual coin-
cide. The problem lies in arriving at the understanding.
For example, in the short run the criminal psychopath seems
likely to obtain more of this world's goods, with less effort, than one who works for a living. Yet if all were
psychopathic in this regard, the consequences to all would
be less satisfactory than at present.

Other issues which must be addressed include the question of which techniques are acceptable and which are not.
Legislation which sets limits on application of behavior
modification techniques has already been passed in some
areas; but have the good techniques been thrown out with
the bad? Another question is whether the techniques themselves have been eliminated or only certain labels for the
procedures by which they are practiced. One area of controversy, for example, is over whether punishment is desirable. Skinner (1953) and to a large degree Ferster,
(Ferster, Culberston and Boren, 1975) seem to believe that
punishment is completely undesirable. Other behavioral psychologists such as Staats (1971) not only advocate it, but
see punishment as essential to effective child rearing.
Which, if any, punishment techniques should be employed:
contingent electric shock, isolation, fines (response cost)?

The final issue is the question of who is to employ the
techniques of behavior control. The most obvious answer is
everyone; in fact this is already the case. But what limits
are to be set on who controls whom? As was true of the ends
toward which control techniques should be applied, so in
the areas of means and agents, the Scripture must be our
standard for right and wrong. The Scripture clearly sanc-
tions parental and governmental agencies; but it also sets
limits with regard to the means and ends toward which these
agencies may control behavior. As London notes, child rearing practices are not only ubiquitous, but universally
deemed desirable. The Scriptures very clearly sanction, in
fact advocate, the training of children. But again, there
are guidelines regarding the methods to be employed.

On the other side of the picture, however, we must be
extremely cautious about coercing people, especially adults,

to do things which we believe to be "for their own good."
Jesus, who claimed to be "the way, the truth, and the life"
did not force his way on anyone. He merely offered it to
those who would accept it. In applying the technology of
behavior control to the human condition, we should keep
this principle prominent in our thinking. Although the
techniques are available, we should restrain our application of them in cases where the individual to whom they would
be applied makes a reasonably informed rejection of their
application. We should also keep in mind that none of the
techniques of behavior control discussed offers the ultimate solution to the problems of mankind.

SOME CONCERNS OF JUSTICE

Allen Verhey
Hope College

"In the great game that is being played, we are the players
as well as being the cards and the stakes."[1] That memorable
figure of de Chardin's says powerfully that the awful responsibility of his own future on earth falls squarely on the
shoulders of man. Dr. London has made us more aware of that
responsibility.

Dr. London has identified two issues as particularly important. The first of these was the Skinnerian view that we
"cannot have a satisfactory social system unless we make
optimum use of the possibilities of behavior control technology."[2] This is, Dr. London says, the heart of the issue.
How are we to understand the crucial words "satisfactory"
and "optimum?" They are obviously evaluative words of one
kind or another. I propose to take them morally, so that the
sentence might be restated, "you cannot have a morally satisfactory social system unless you make morally reasonable use
of behavior control technology." That may or may not be what
is meant, but such a moral interpretation does, I think, get
us close to the heart of the issue. It remains to discern
what a morally reasonable use might be.

The second statement Dr. London emphasized is not a puzzle
at all, but the corner-piece of any constructive attempt to
deal with the difficult issue of exactly what a morally reasonable use might be. "The value issues," Dr. London says,
"which require external legislation and regulation are those
in which the technology involved enables some people to exercise control over others."[3] He identifies the technology of

ALLEN VERHEY earned a B.D. from Calvin Tehological Seminary
in 1969 a.id a Ph.D. in Religious Studies from Yale University
in 1975. His specialties are Theological Ethics and New Testament Studies and their interrelation. Dr. Verhey has taught
at Calvin Seminary and is currently Assistant Professor of
Religion and Bible at Hope College, Holland, Michigan. He has
taught courses in medical ethics to seminarians, medical professionals, and college undergraduates.

215

behavior control with other forms of social power. Further, Dr. London says in another place, "Technologies do not create or answer moral problems; only men do that. The final issues of moral intercourse, accordingly, do not depend on how men are able to use their tools, but on how they are willing to use each other. The moral problem of behavior control is the problem of how to use power justly."[4] Perhaps, then, we may presume to restate London's statement with more precision: "you cannot have a morally acceptable social system unless you use the possibilities of behavior control technology justly (and only justly)." That is, I think, the very heart of the issue. And that, of course, forces the question of justice.

Unfortunately, Dr. London resists a straightforward and candid address to the question of justice. He approaches the issue helpfully from many sides: the importance of maximizing a person's sense of freedom, the priority of the individual's rights over the social good, and interesting remarks on the history of political theory and practice. These are important, and I express my general appreciation of them. Nevertheless, a more straightforward address to the question of justice seems necessary. The recent work of John Rawls may be helpful.

PRINCIPLES OF JUSTICE

John Rawls postulates, for the sake of analysis, a state of nature in which there are no rich and no poor, no black and no white, no distinctions of any kind. We might legitimately add then, that in this state of nature there are no "mentally ill" and no "normal", no behavior controllers and no behavior-controlled. These people must together decide on certain fundamental rules for the conduct of their life together when they "fall" from this state of nature. Then some will be rich, but who? And some will be black, but who? And some will have access to the power of behavior control, but who? In this situation Rawls argues all would agree that their life together must be governed by these principles of justice: "First, each person is to have an equal right to the most extensive basic liberty compatible with a similar liberty for others. Second, social and economic inequalities are to be arranged so that they are both (a) reasonably expected to be to everyone's advantage, and (b) attached to positions and offices open to all."[5] These principles, which can be shorthanded as maximum freedom and presumptive equality, both license and limit social power. Each has a right to equal freedom. That's basic. Inequalities in power, whether political power or economic power or the power of behavior

216

control technology, are legitimate if and only if such inequalities are to everyone's approximately equal advantage
and attached to positions and offices open to all.

We all know the mythology developed by *laissez faire* cap-
italism in its attempts to justify itself before criteria
very much like these. Some invisible hand guides the market
place so that free exchange works for the advantage of all
without unacceptable inequities. The road from rags to riches
is supposedly open to all. In the future we must be wary of
such mythologies designed by the powerful to pay lip service
to these principles, but more about that later. The point
now is simply to recommend our use of these principles of
justice in our attempt to deal with the legitimacy of the
power of behavior control technology and of its use of its
power.

My first observation about justice is that Dr. London has
not done us a service by suggesting that the "lower limits"
of conditioning are unchallengable. He himself saw such pervasive conditioning as "promising for the resolution of the
value problem." But his blanket endorsement under the cover
of toilet-training is no resolution; it's a whitewash! I do
quite agree that there is a spectrum of social power, as
London's term "lower limits" implies. A look at such a spectrum may be helpful.

I suggest that the spectrum extends from cooperation
through intervention to aggression.[6] On the cooperation end
of the spectrum x gets y to do z by discourse, giving him
information and reasons. Here x regards y's will and his
consent as important. X gives y the opportunity to dissent
and to give his own reasons. Neither x's will nor y's will
is dominant; rather, their mutual regard for each other and
for the truth takes precedence over both. They stand as
equals. Even in the failure to agree, they respect and regard each other as equals. Such cooperation is the very
model of justice.

On the aggression side of the spectrum x gets y to do z
by violence, whether bodily, psychologically, or by other
forms of violence. Here x does not regard y's will. X ar-
bitraritly disregards y's reasons. X's will and x's reasons
are decisive. They stand as unequal, with x in arbitrary
dominance over y. Arbitrary dominance is the very model of
injustice.

Between cooperation and arbitrary dominance lies the morally significant range we have called intervention. Here,
close to cooperation, at the "lower limits", lies propaganda.

217

X gets y to do z by quasi-rational persuasion. Here x leaves y with a sense of reason - hearing and choosing, but a mutual regard for each other and the truth no longer takes precedence over x's will. To the degree that propaganda approaches truthful information it approaches cooperation. As truthfulness declines, it approaches arbitrary dominance. Advertising, for example, is a form of propaganda. It can be ranked on the spectrum according to whether it gives bona fide information (here it shades into cooperation) or uses "hidden persuaders" (here it shades into arbitrary dominance).

Parenthetically, I would express my discomfort with Dr. London's use of "a sense of freedom" rather than simply "freedom". The propagandist protects my sense of being a rational agent but does not respect my rational agency. The terrifying thing about behavior control technology is its ability to preserve my sense of freedom while it effectively destroys any genuine freedom. It promises (or threatens) not just to get me to do z but to get me to want to do z.

But I must return to the point I am really trying to make. It is this: All intervention, all departures from the model of cooperation, require justification. Propaganda, advertising, promises, threats, rewards, punishments, whether at the "lower limits" or not, calls out for justification. It must be stated that the point is not that no intervention is ever justified and not that all intervention is necessarily arbitrary dominance. The point is rather that it must be justified. It can be justified morally only in or as hindering hindrances to freedom (to use an old Kantian formulation of just power).[7]

Even the power relation of parents and children needs justification.[8] As a parent I can testify that the politics of my parenting sometimes escalates from forms of cooperation to quasi-rational persuasion to promises to threats to coercion fairly quickly. In fact, sometimes it escalates so quickly as to raise questions of arbitrary dominance in the mind of my seven-year-old. The justice of my parenting does not receive a blanket endorsement because I value toilet-training. My parenting is justifiable if and only if it enables my son to enter cooperative relationships when he is older.[9]

To summarize the first point under justice: The "lower limits" of conditioning are challengable. We may not be less than morally and politically vigilant wherever intervention takes place, even at the "lower limits". All departures from freedom and equity demand justification. Some of them are justifiable - but only by the principles of

218

maximum freedom and presumptive equality.[10]

The second observation under the rubric of justice turns from the ways in which intervention already characterizes our lives to long range effects. The future may not be quite the gentle utopia London apparently predicts without enthusiastically hoping for. The reason for a more cautious disposition toward the future introduces the importance of technology.[11]

Technology has a self-stimulating impetus, London admits, but he says all its uses are channeled by the value systems of the technicians, their mentors, and their customers. Moreover, their development is a simple function of the value system of the society. Such a view of technology is popular. It is simply a way of getting what we already want. It's the way to get to utopia, the millenium. One could list the utopian promises of the technologists, but that would take too long and is not a very fruitful activity. The gentle prophets of the technological millenium are, of course, aware of ecological threats and other unhappy products of technology. However, they usually simply add an answer to such problems to the already impressive (not to say incredible) roster of technology's promises. Indeed, that gets us close to the fundamental principle of technological utopians: Technological innovation exhibits a tendency to affect the general welfare over the long haul. *Laissez faire*! The marketplace will guide technology to great human betterment. The invisible hand can be trusted after all!

That is not to say that technology has been insignificant and unappreciated. Of course it has been. None of us, as London says, are willing to do without toilets, let alone toilet training. It is to say that we ought to be a good deal more circumspect and cautious about technology. We ought not to trust some invisible hand over the marketplace. We should not accept tempting new mythologies which promise what we want while they pay lip service to freedom and equity.

TECHNOLOGY, JUSTICE AND DEMOCRACY

The point was to focus, I promised, on justice. And this is it: Technology once contributed to the development of a democratic ethos in Europe and North America, but it now challenges it and erodes it. It has done so partially by providing room for deviancy, as London suggested. But it contributed more fundamentally to the destruction of the social gap between the masses and the ruling classes.

219

The printing press encouraged the spread of literacy. That supported the ordinary man's reading of the Bible, and Protestant Christianity was able to challenge the arbitrary dominance of hierarchy. That literacy was sufficient to support a growing printing technology; together they supported newspapers and political tracts. The ordinary man was thus able to challenge the arbitrary power of the ruling class.

Roads, postal service, the invigoration of towns by new tools and new merchants, all contributed to make the social experience of the ruling classes. The ordinary man gained experience in balancing present social means with distant social goals in ways not unlike the business of the rulers. All this demystified the position and power of princes and priests and aided the transition from traditional authority to rational-legal authority (to use Max Weber's categories).

Technology contributed therefore to the democratic ethos. This is one of the things that has made it a boon to man. But now, I suggested, its role is shifting. It is morally necessary now to ask whether the new technology does not seriously threaten the democratic ethos. I believe it erodes it in at least two ways: illiteracy and mystification. There is a decline in the kind of literacy that gives understanding of and access to the political and social character of the new technology and the power it holds. The growing gap between technical languages and the language of ordinary men signals the erosion of what one may take to be essential to the democratic ethos.

No less important is the mystification of the technocrat. When the awesome mystery with which ordinary man regards science is given to the new controllers, the mystique of former princes and priests seems prosaic. There is a new gap in social experience, the gap between those who know how to get things done and ordinary men. These tendencies of the new technology erode the democratic ethos and threaten a new elitism.

In fact, London does not seem to deny the emergence of a new elite. But such an elite wil presumably care for and respect the masses under their control. Such an elite is expected to be creative and hardworking and altruistic while the ordinary man engages in purely hedonistic preoccupations. But isn't it fair to ask just how much respect we carefree pleasure lovers will get from the industrious controllers? Their altruism will have to bear a heavy load. I would rather trust justice. We would better work for the preservation of a democratic ethos. It is unrealistic to make such demands on the gentle controller's altruism in a world where we do not keep the command to love and are prone to hate God and our neighbor (cf. Heidelberg Catechism, Lord's Day 2). We must put the burden of proof for any form of intervention on the controller now in order to avoid such intolerable demands on the benevolent conditioner's charity later.

Concretely, we can trust justice and preserve a democratic ethos by restricting behavior control technology, excepting certain specifiable classes of clearly licensed uses of this power, and giving discretionary power to civilian review boards (which include those to be controlled), for more ambiguous uses. A certain legal casuistry is perhaps inevitable here, but it is better than trusting our future to the mythology of laissez faire technology and the altruism of the gentle technologically elite.

FOOTNOTES

[1]Teilhard de Chardin, *The Phenomenen of Man* (New York: Harper & Row, 1965), p. 230.

[2]Perry London, "Behavior Control, Values and the Future," p. 193.

[3]Perry London, *ibid.*

[4]Perry London, *Behavior Control* (New York: Harper & Row, 1969), P. 199.

[5]John Rawls, *A Theory of Justice* (Cambridge: Harvard University Press, 1971).

[6]Cf. David Little's significant address to foreign policy decisions in his "Reason, Rationalization, and Foreign Policy," for the Council on Religion and International Affairs, p. 21. The just war tradition to which he refers could be very helpful in the adjudication of questions in medical ethics raised by the new resources of power.

[7]Cf. I. Kant, *Metaphysical Elements of Justice* (New York: Bobbs-Merrill, 1965).

[8]Some relief from the typically sentimental analysis of parenting can be found in the recent bood of Sidney Cornelis Callahan, *Parenting: Principles and Politics of Parenthood* (New York: Pengwin Books, 1974).

221

[9]The work of Jean Piaget, *The Moral Judgment of the Child* (New York: Free Press, 1965), makes a good deal of the development of the capacity for cooperation in children.

[10]An example of a use of behavior control which is clearly justifiable by its enabling cooperation is the work Dr. London mentioned might be possible with autistic children.

[11]The following analysis of technology is indebted to John Mc Dermott, "Technology: The Opiate of the Intellectuals," *The New York Review of Books, XIII #2* (July 31, 1969).

BEHAVIORAL ENGINEERING AND SPIRITUAL DEVELOPMENT

Paul W. Clement
Fuller Graduate School of Psychology

The safest thing for the reader will be not to believe a
single word that follows. Hopefully, however, some individuals will respond by exploring the possibilities of developing
a psychological engineering for spiritual development.

GENERAL RESPONSES

After going through Dr. London's paper several times, I
discovered that I disagreed with eight of his points, wished
that three additional points had been made, and agreed with
the rest.

Most of the items of active disagreement involved minor
issues. For example, I don't believe there is such a thing
as a "...safe means of committing suicide."

Indirect disagreements occurred when Dr. London didn't say
something which I thought needed to be said at a particular
point. For example, he asserted that: (a) the availability of
a technology leads to a search for uses, and (b) values may
foster a technology. What was not stated was that there is no
necessary or logical relationship between a given technology
and a given value system. Often people of a particular value
system reject a particular technology because the originator
of that technology does not hold the value system of the lis-
tener or reader.

PAUL W. CLEMENT is Director of The Psychological Center and
Professor in the Graduate School of Psychology at Fuller Theological Seminary. He is a practicing clinical psychologist,
a Diplomate of the American Board of Professional Psychology,
a past president of the California State Psychological Association, and a contributing editor to the JOURNAL OF PROFESSIONAL
PSYCHOLOGY. Dr. Clement's research and writing have focused
on applying concepts and procedures from general psychology to
the solution of behavioral problems in children. He has published over 35 atricles, two films and one tape series on behavior modification with children.

I fully agree with Dr. London's three major theses: (a) Non-coercive controls are most important and desirable; (b) Maximizing volition or choice is what is not important; (c) The ultimate society will "...involve a constant tension between individual needs and the requirements of society." Looking at the third statement, there is no guarantee that an adequate balance will be achieved between personal and societal needs. All parties involved have the option, however, of working to maintain balance in the face of the dynamic tensions which confront the individual, group, or society that attempts to walk the fence of moral behavior.

In contrast, some options simply do not exist. A case in point is that "...we cannot forego behavior control by conditioning." I agree with this assumption. Conditioning is going on all of the time. It is not new. What is new, as of the twentieth century, is the application of scientific methodology to the study of conditioning process, and to the advancement of human welfare. In the past, if not the present, most people have assumed that science can not be applied to human behavior.

Science is involved with discovering and controlling empirically verifiable events through observing, categorizing, systematizing, predicting, and manipulating the phenomena of concern. The development of prediction and control is usually followed by the development of a technology.

Strictly speaking, scientists don't invent anything. They only observe what already exists and try out combinations of pre-existing components. Such a point of view is certainly compatible with the Scriptures: "What has been is what will be, and what has been done is what will be done; and there is nothing new under the sun" (Ecclesiastes 1:9).

BEHAVIOR CONTROL

I assume that Dr. London and I would be able to agree on a definition of "behavior." Behavior includes reflexes, emotions, perception, and sensations, actions, physiological events controlled via biofeedback, and thoughts.

In order to develop the psychological type of human engineering, a functional analysis of behavior must be performed. The pivotal question is "How can a given behavior in a given setting be strengthened, weakened, maintained, or produced, or eliminated?"

Unfortunately, there are many people who do not understand

the difference between the behavior studied by Pavlov[1] and that studied by Skinner.[2] In the case of the behavior studied by Pavlov, our world (the environment) produces changes in our behavior: the world changes first and our behavior changes second. Reflexes, emotions, perceptions, and sensations are such behaviors.

In the case of the behavior studied by Skinner, our behavior produces a change in our world (the environment): our behavior changes first and the world changes second. Actions, physiological events controlled via biofeedback, and thoughts are examples.

Educational or therapeutic procedures based upon Pavlovian or Skinnerian conditioning have been developed extensively during the last 15 years. These procedures are often powerful, but they are relatively slow when compared to a third type of conditioning. The latter is most prevalent in normal human beings from late infancy onward.

Although this does not seem to be the case with infra-human species, man can learn (almost) anything which can be taught via Pavlovian or Skinnerian techniques more rapidly through observational learning (modeling). Observational learning is an extension of both Pavlovian and Skinnerian conditioning. It has been extensively researched by Albert Bandura[3] at Stanford University. A behavioral technology based upon observational learning is rapidly developing.

The research and writing of Pavlov, Skinner, Bandura, and many other behaviorists has stimulated the growth of an extensive psychological technology that is available to us in the church, if we want to use it. Most Christians seem to have assumed that behavioral psychology has nothing to offer the church or individual believers. Such is not the case. Just as behavioral psychology has had a significant impact on clinical psychology, psychiatry, psychotherapy, and education[4,5,6,7], it can have a great positive impact on the church.

One of the newer developments within behavioral psychology has been an emphasis on mechanisms of self-control,[8,9] self-management,[10,11] and self-regulation.[12,13] This literature makes clear that one of the most exciting things about people is their ability to be both experimenter and subject, therapist and patient, counselor and client within the same skin. To the best of my knowledge, there is nothing that a behavioral engineer could do to me that I could not do to myself. I am, therefore, assuming that this growing technology of self-control will make all of the specific conditioning strategies of the behavioral psychologist acceptable to the informed Christian.

VALUES

Technology only tells us how to do something. Although behavioral technology may tell us how to change behavior, it doesn't give the slightest clue regarding in what direction to change behavior or which behaviors should be changed. Technology says nothing about the goals that we want to or should pursue; it doesn't deal with the issue of values. Values need to come from some other source.

In looking at Dr. London's paper as it deals with values, it is descriptive; that is, it describes what appear to be moral issues in society today. It is not prescriptive. Dr. London did not attempt to tell us what our values or goals should be. In assuming such a posture he is congruent with the stance which has been assumed by the American Psychological Association.

Psychologists have published a large number of papers and documents on the ethical issues involved in professional practice and scientific research. The two most significant such documents are the CASEBOOK ON ETHICAL STANDARDS OF PSYCHOLOGISTS[14] and ETHICAL PRINCIPLES IN THE CONDUCT OF RESEARCH WITH HUMAN PARTICIPANTS.[15] These publications clearly indicate what psychologists tell themselves is appropriate and inappropriate. Many additional journal articles have appeared in recent years dealing with the ethical implications of behavioral engineering, in particular. One such paper, "Behavior Therapy and Civil Liberties"[16] appeared just a few days before the International Conference on Human Engineering and the Future of Man began. These prominent behavior therapists tried to give some very concrete guidelines for behavioral researchers. Of all the behaviorally oriented psychologists, Dr. London has probably written the most on the ethical/moral issues of behavioral engineering.

All of the above publications focus on *process* values. To a great extent process values deal with certain aspects of how we change behavior. For example, two fundamental process values are: (a) Patients or research subjects should be informed about the procedures to which they will be exposed, including the possible risks and benefits of undergoing such procedures; and (b) The situation should be free of coercive elements. These are highly compatible with, and probably stem from, a number of biblical teachings, such as "...when the Holy Spirit controls our lives He will produce this kind of fruit in us: love, joy, peace, patience, kindness, goodness, faithfulness, gentleness, and self-control" (Galatians 5:22-23).

Process values provide guidelines on the relationship between the therapist and the patient, the investigator and the subject, the controller and the controllee. They do not provide any guidance as to who should be identified as a patient, who should be selected as a research subject, who should be controlled, what behaviors should be changed in what direction, etc. Such questions fall outside the realm of psychology as a discipline, though they are the very questions which each psychologist must personally answer.

Outcome values are often hard to handle – perhaps because they centrally involve human passions. Technology and process values are rather sterile until they are fertilized by a compelling emotion, i.e., a passion. The choice of objects (targets) for such intense emotional drive involves outcome values. These values often appear to determine our involvement in love, war, politics, religion, and vocation. A person who only has a technology and process values is dead. To be alive requires having a technology, plus process values plus outcome values (passion).

For the Christian the Scriptures are the most important source of outcome values. The Bible has a number of suggestions on what the objects of our passions ought to be. For example, "...you must love the Lord your God with all your heart, and with all your soul, and with all your strength, and with all your mind. And you must love your neighbor just as much as you love yourself" (Luke 10:27). Another passage which presents a distinctively Christian outcome value is "Jesus' disciples saw Him do many other miracles besides the ones told about in this book, but these are recorded so that you will believe that He is the Messiah, the Son of God, and that believing in Him you will have life" (John 20:30-31).

I think that there is something available within behavioral technology to help a person be congruent with his process values and move toward his outcome values. My concern is that Christians, in particular, will: (a) confuse outcome with process values; (b) confuse goals with the means used for pursuing those goals; (c) confuse philosophy with technology; (d) confuse an action with a motive; and (e) confuse theology with the means of communicating the Gospel. That such confusion has existed throughout the history of the church is suggested by the following passage:

227

Some, of course, are preaching the Good News
because they are jealous of the way God has used
me. They want reputations as fearless preachers!
But others have purer motives, preaching because
they love me, for they know that the Lord has
brought me here to use me to defend the Truth.
And some preach to make me jealous, thinking
that their success will add to my sorrows here
in jail! But whatever their motive for doing it,
the fact remains that the Good News about Christ
is being preached and I am glad. (Philippians
1:15-17).

SPIRITUAL DEVELOPMENT

Some people are troubled by the idea of precise, planned,
behavioral interventions at any level for any reason. The
sense of threat goes up as we discuss behavioral control of
personal spiritual development. The assumption seems to be
that if one "engineers" spiritual development, there can be
no spiritual development. I don't believe this is the case.

I think that in its history the church has provided some
tremendous examples of human engineering. Unfortunately,
at some points it may have violated some of our process values. Nevertheless, some of our very best examples of behavioral engineering can be found in church history. The problem has been that Christians haven't explored the specific
mechanisms of behavioral control. We have assumed that human
behavior can not be analyzed in such a way as to develop an
improved psychological technology for spreading the Gospel.
The only new thought being presented here is that we need to
carry out a systematic analysis of what leads to what. Namely,
we have the opportunity to experimentally analyze what psychological events will best lead us toward the goals (both
process and outcome) which we, as Christians, would want to
promote. Secondly, we will need to identify those psychological events which hinder the pursuit of Christian values.

I do not believe that the application of behavioral technology will dehumanize man. Man in all cases remains human.
In no case is he merely a machine, because he is alive. What
being alive means is a key issue.

Perhaps two of Skinner's rats can shed some light on the
problem of what it means to be alive. You may have seen a
cartoon of these two rats. They live in two adjacent Skinner
boxes (operant conditioning chambers). The walls are transparent and perforated, so that they can see each other and
communicate. Each box is simply furnished with a bed, one

light, a lever, a food dispenser, and a water bottle. One rat appears to be happy. He smiles, wags his tail, licks his chops, and eats his food pellets with gusto. When he finishes one pellet of food, he presses the lever until another drops out of the dispenser.

The second rat appears to be sad. He is tearful, doesn't wag his tail, mopes around his cage, and gives great sighs between nibbles on his pellet of food. When he finishes one pellet of food, he eventually drags himself over to the lever, presses it, and gets another pellet.

Both of these little rodents are philosophical rats. They spend much time debating on the meaning of life. Their dialogue typically runs as follows:

Sad Rat: (Turning toward the happy rat.) "What on earth are you always smiling about?"
Happy Rat: "What do you mean?"
Sad Rat: "Well, you dummy, we are trapped inside these dumb Skinner boxes. Everything we do is under Skinner's control."
Happy Rat: (Looking surprised.) "I don't understand. I've got control of Skinner."
Sad Rat: "You're a jerk! Skinner has programmed everything that happens to you. Your situation is hopeless."
Happy Rat: (Demonstrating.) "You're wrong. I control Skinner. I can get him to give me food whenever I want it. Watch this!" (He presses the lever and gets a pellet of food.)

These rats will spend the rest of their lives arguing over who has really got control. There is no question in my mind, however, as to who is the more functional rat, and which rat is living more "in the Spirit." I would want to join the happy rat and help him find other ways of controlling Skinner, who in turn, will control the rat, who will control Skinner, *ad infinitum.*

Any satisfactory situation involving behavioral engineering will be one which allows for two-way control. The relationships need to be reciprocal. Wherever there is control, there must be counter control. If we want to teach, train, treat, or evangelize someone, we must provide a system which allows the "learner" to influence our behavior in return. From the perspective of process values, things go wrong when the learner does not have an adequate way of influencing how the teaching is proceeding.

From the perspective of outcome values, things go wrong

when the teacher and learner disagree as to what shall be learned. When we allow for a process in which the teacher tells the learner what he wants to teach, dialogue can occur. Such dialogue can form the basis for the learner and teacher signing a contract which is mutually acceptable and which is free of coercion. The idea of contracting is basic to most clinical uses of behavioral technology today. Since "contracting" is a synonym for "covenanting" and covenanting is a well-established biblical concept, Christians should not have to avoid behavioral technology.

But, the reader may protest, "Clement, you have tried to replace the work of the Holy Spirit by your behavioral technology!" Such a protestation misses a fundamental point. The Holy Spirit provides guidance about goals (i.e., as to what behaviors are desired) without necessarily showing specifically how to reach these goals. Behavioral technology can provide much of the how. When Christians begin to realize the potential of behavioral technology for doing good and for being carried into the church and religious community, we will have moved into an exciting area that some are calling "behavior modification of the spirit."[17]

FOOTNOTES

[1]I.P. Pavlov, *Selected Works*, trans. S.P. Belsky (Moscow: Foreign Languages Publishing House, 1955), (contains papers originally published during the first three and a half decades of the twentieth century).

[2]B.F. Skinner, *Cumulative Record: A Selection of Papers*, 3rd. ed. (New York: Appleton-Century-Croft, 1972).

[3]A. Bandura, ed., *Psychological Modeling: Conflicting Theories* (Chicago: Aldine-Atherton, 1971).

[4]A. Bandura, *Principles of Behavior Modification* (New York: Holt, Rinehart & Winston, 1969).

[5]F.H. Kanfer & J.S. Phillips, *Learning Foundations of Behavior Therapy* (New York: John Wiley & Sons, 1970).

[6]A.A. Lazarus, *Behavior Therapy & Beyond* (New York: McGraw-Hill, 1971).

230

[7]J. Wolpe, *The Practice of Behavior Therapy*, 2nd ed., (New York: Pergamon Press, 1973).

[8]C. Foster, *Developing Self-Control* (Kalamazoo, Michigan: Behaviordelia, Inc., 1974).

[9]N.R. Goldfried & M. Merbaum, *Behavior Change through Self-Control* (New York: Holt, Rinehart & Winston, Inc., 1973).

[10]R.L. Williams & J.D. Long, *Toward a Self-managed Lifestyle* (Boston: Houghton Mifflin Co., 1975).

[11]C.E. Thoresen & M.J. Mahoney, *Behavioral Self-Control* (New York: Holt, Rinehart & Winston, 1974).

[12]D.L. Watson & R.G. Tharp, *Self-directed Behavior: Self-Modification for Personal Adjustment* (Belmont, Calif.: Brooks-Cole Publishing Co., 1972).

[13]M.J. Mahoney & C.E. Thorsen, *Self-Control: Power to the Person* (Monterey, Calif.: Brooks-Cole Publishing Co., 1974).

[14]*Case Book on Ethical Standards of Psychologists* (Washington, D.C.: American Psychological Association, 1965).

[15]*Ethical Principles in the Conduct of Research with Human Participants* (Washington, D.C.: American Psychological Association, 1973).

[16]G.C. Davison, & R.E. Stuart, "Behavior Therapy and Civil Liberties," *American Psychologist,* (1975), 30, pp. 754-763.

[17]P.W. Clement, "Behavior Modification of the Spirit," Keynote address presented at the Western Association of Christians for Psychological Studies, (Santa Barbara, Calif., May, 1974).

PART VII
SCIENCE, PUBLIC POLICY, AND THE CHURCH

Mark Hatfield instructively reflects upon three historical episodes that involved the interaction of the
religious, political and scientific communities. He asserts
that the Christian community must develop a view of the
state that includes the possibility of fundamental dissent if
the state violates basic moral guidelines. The church must
be a critic and not a naive partner. Secondly, Senator Hatfield
reviews the historical and current involvement of the
government in biomedical and behavioral research. Finally,
Hatfield cautions the Christian community to avoid cliquishness and arrogance by becoming actively involved with all
others concerned about the implications of human engineering.
He suggests that such involvement will be much more salient
to both the scientific and legislative communities than
statements out of an isolationist corner.

Sociologist John Scanzoni picks up on these last concerns
and reminds the evangelical community that others have
already been intimately involved in these concerns for much
longer. He encourages the Christian community to overcome
its fear and suspicion, and to work together with those of
different theological and value orientations to arrive at a
mutually agreeable set of guidelines. Scanzoni points out
that, because of the way in which Christianity has pervaded
the value system of the United States, evangelicals may not
have consistently unique contributions to make. Nevertheless they should be uniquely motivated to involvement in
this area of concern. Finally, Dr. Scanzoni argues the need
for the general Christian community to challenge young
people to become involved in science and ethics and to become
participants in wrestling with the complex issues of our
day.

John Olthuis brings the response of a lawyer to Senator
Hatfield's address. He focuses on the application of
biblical ideas to the nature and operation of the state.
Olthuis asserts that the responsibility of the state is to

233

provide a framework which encourages individual development in line with God's commandments of love. He then explores the meaning of those commandments with regard to justice and response to authority. Finally, Olthuis criticizes the current functioning of North American governments and argues that technology has come to be seen by the government as a way to deliver happiness. In this context, abusive application of human engineering knowledge is likely, unless Christians become politically involved.

Finally, noted theologian Carl Henry focuses upon the difficulty of sustaining democratic processes and values in the face of the swift decisions that current and future technology have forced upon decision-makers. He points out the high probability that the issue of public interest and national security will likely be a crucial concern in the future controversies that arise about the justification of human experimentation without informed consent. Finally, Dr. Henry challenges the Christian community to become actively involved in the political and decision-making processes. He believes that Christians have unique contributions to make to the way in which science and ethics are conceptualized because of fundamental presuppositional and value differences.

PUBLIC POLICY AND HUMAN ENGINEERING

Senator Mark O. Hatfield

I want to begin my presentation with some reflections from
several historical episodes, then consider the Christian view
of the state. I will continue with an overview of what is
being done now by the federal government in the field of human engineering and conclude by making some suggestions for
the expression of Christian ethics and concern.

The three episodes from the past which come to mind at this
time could be symbolized by three objects: a monkey, a geranium, and a cocktail. While you puzzle about the latter two,
let me begin with the monkey, the silent participant in the
trial of J.T. Scopes in Dayton, Tennessee.

The basic issue in the Scopes trial resembles one which
faces you this week: the role of the state and of Christians
in responding to new developments in science. As the historian
Robert Linder has explained so well,[1] William Jennings Bryan
has been badly abused by historians. They have allowed the
brilliance of his early career to be overshadowed by his weak
performance against Clarence Darrow. Far from being a narrow-
minded, backward, inarticulate politician-turned-crusader,
Bryan during his prime years set an excellent example of
Christian involvement in politics. He effectevely dealt with
the full range of social concerns.

But the mistake of Bryan and the fundamentalist Christians
he represented in the Scopes trial should be noted very

HON. MARK O. HATFIELD is a United States Senator, Oregon,
and is currently serving his second term of office. He is a
member of the Senate Appropriations and Senate Interior/Insular
Affairs Committees, and Ranking Minority Member of the Senate
Rules Committee. Prior to election to the Senate he served
as Governor of Oregon and Associate Professor of Political Science at Willamette University in Salem, Oregon.

carefully by those applying Christian ethics to science.
Bryan and his fellow believers failed to understand how
devastating the press could be against someone who would
defy the new cult of science. Bryan was no longer able
intellectually and emotionally to relate to the reporters
any more than he was to the brilliant Clarence Darrow.
From that point his logical fallacies and ineptness began
to build a negative image for evangelicals, which must be
corrected, not renewed.

I am confident that the failings of the past will not be
repeated. Today, evangelicals are not represented by poorly-
informed, inarticulate, unsophisticated and reactionary scientists and attorneys. Indeed, William Jennings Bryan did
not deserve the caricature. Some will, however, still expect
this of evangelicals. Christians must be very careful about
formulating statements or making public proclamations before
they have given the utmost consideration to the issues. They
also must be careful about allowing the discussion to be
diverted to indefensible ground. Bryan did not lose his
fight on the issue of creation versus evolution. Rather,
it was lost on the issue of freedom of inquiry and education
versus repression and narrowness.

The second episode is symbolized by a potted geranium.
It sat in a window of a small, bare room in a children's
ward in a hospital called Eglfing-Haar in wartime Germany.
The flower was carefully watered and remained as a strangely
discordant note of beauty in a setting of gross violence and
degradation. The small room was used to administer fatal
injections to selected children from the hospital. This was
the very nadir of ethics in science.[2]

The term "euthanasia" was applied to this program of
eliminating the retarded, the handicapped, the diseased and
the maimed. The objects ranged from the youngest children
to the senile and elderly. Many of the estimated 275,000
psychiatric patients killed were simply exterminated in
gas chambers or with lethal injections. The primary significance for our discussion is the unbelieveable relationship between the scientists, the government and the
church.

Since we can blame Hitler and his major subordinates for
the extermination of the Jews, we might suppose that they
also conceived and directed the so-called euthanasia program.
This was not the case, according to evidence from the post-war
trials. Long before the Nazis came into power, psychiatrists

laid the groundwork for the idea of developing a master race by eliminating undesirables. It was a prominent professor of psychiatry and an equally eminent professor of law who wrote the book, THE RELEASE OF THE DESTRUCTION OF LIFE DEVOID OF VALUE in 1920, from which much of the justification came. Beginning with an exaggerated notion of the place of heredity in mental disorders, these professors applied their ideas to a mandatory sterilization program and then to the most blatant of extermination efforts.

An estimated fifty psychiatrists took part in this mass murder in the name of enlightened science. Among them were a dozen full professors of the best German universities and others whose scholarly work is still remembered. These psychiatrists obtained Hitler's approval after they had begun the program, but apparently not at his suggestion. They conducted their work voluntarily and some continued to see nothing wrong with it during the trials. They met together as scholars and administrators periodically, just as you are meeting here. They had no apparent remorse over their actions. One exception was a professor of chemistry who challenged the program. He said, "The patient is no longer a human being needing help, but merely an object whose value is measured according to whether his life or destruction is more expedient for the nation." Unfortunately, he was not heard.

What went wrong with the values of the scientists in Nazi Germany? Of greatest relevance to our theme is the fact that the scientists were completely without any effective restraint. Those in power in the state certainly did not object to euthanasia. Though it may not have been their idea, they had no reason to oppose it. More importantly, the church completely failed to provide a conscience for the scientists and politicians. The clergymen did deliver sermons and wrote letters to the officials. They also sought to exempt the religious sisters from assisting in the killing and objected to the transfer of patients to the hospitals where the lives were taken. This effort had no demonstrable effect. It is strikingly similar to some of the efforts of churchmen today in attempting to influence political policy.

A poignant episode took place as a part of this mild religious resistance. A pastor, Fritz von Bodelschwingh, was so troubled by what he had learned about euthanasia that he had a long conference with Dr. Karl Brandt, one of the psychiatrists who directed the program. Brandt was a friend and admirer of Albert Schweitzer, even then a great spokesman and example of reverence for life. But even Brandt could see no reason for halting the deaths, and the program continued until the end of the war.

The third episode involves a cocktail being served after dinner a few miles out of the nation's capitol. It was 1953. The United States had barely recovered from the irrational fears of McCarthyism and had plunged into another war. The drinks were being shared by several scientists at a Virginia retreat. Among them was Frank Olson, one of the civilian researchers who worked at Fort Detrick in Maryland. The bulk of the research involved bacteriology as applied to defense - "germ warfare" in layman's language. As brought out by the recent Rockefeller Commission report, and in subsequent new investigations, the sponsor of the research contract, the Central Intelligence Agency, was interested in research quite unrelated to bacteria and viruses. The cocktails that night in 1953 contained LSD, something which was only revealed to the five scientists after they had consumed the substance.[3]

Of course, the outcome of the "experiment" was tragic. Frank Olson underwent a dramatic change of behavior, to the point that CIA agents felt compelled to take him to a psychiatrist in New York. There he plunged to his death from a tenth floor window. The CIA told Olson's family he had taken his own life and did not divulge the link to the LSD experiment.

What kind of relationship was there among the scientists, the state and the church in the CIA episode? While only a few of the scientists may have dealt with the LSD research, most were experimenting with sinister tools for use against some enemy. These included germs, gas and even a chemical which could carry LSD to an entire enemy force, as well as civilians. The Army was also engaged very extensively in this work.

We have to ask whether scientists can ethically use their skills for these ends. Of course the government must be blamed for tolerating an agency which could secretly sponsor such research; Congressional and executive control over the CIA has been minimal. We must soberly ask what kind of human engineering and destructive techniques are even now being tested by the CIA and the Department of Defense.

We should note that Christians and citizens in general did not protest the research at the time. They still have not

to any great extent. Twenty-two years ago most of us
had no idea what LSD was, much less any notion that it was
being used in experimental behavior modification.

TOWARD A CHRISTIAN VIEW OF THE STATE

I selected these particular episodes because they force
us to answer some weighty questions. What is the Christian
view of the state? How do Christian values apply to the
state and to bioethics? What is the state doing now, and
what should it be doing in regard to human engineering?

In spite of the contradictory interpretations of the
nature of the state which have been put forth by Christians,
I see in the Bible a clear portrayal of the state. It is
presented as an imperfect, but necessary, means of achieving
order in society. Paul writes in Ephesians 1:18-23:

I pray that your inward eyes may be illumined, so
that you may know what is the hope to which He calls
you, what the wealth and glory of the share He offers
you among His people in their heritage, and how vast
the resources of His power open to us who trust in
Him. They are measured by His strength and might
which He exerted in Christ when He raised Him from
the dead, when He enthroned Him at His right hand in
the heavenly realms, far above all government and
sovereignty that can be named, not only in this age
but in the age to come. He put everything in subjection beneath His feet, and appointed Him as supreme
head to the Church, which is His body and as such
holds within it the fullness of Him who Himself
receives the entire fullness of God.

The Apostle was speaking not just of political powers in
a limited sense, but of all the ideologies, forces and institutions which make up society. Paul was saying that Christ
had unmasked and disarmed the authority of these cultural
forces. And to those at Galatia, Paul wrote:

Now that you have come to know God, or rather to
be known by God, how can you then turn again to the
weak and beggardly world powers to whom you want to
be enslaved once more? (Galatians 4:9).

Thus, the message of the New Testament is that Jesus Christ
is Lord. In a fallen world awaiting its final redemption,

the "powers" are ultimately used by a Sovereign God for His purposes. Yet the Christian has been set free from their dominion and authority through the Lordship of the resurrected Christ. The state, then, has a provisional and temporary role, but a legitimate and important one in maintaining coherence in human society.

The problem has been that Christians have tended to take a few verses from Scripture to justify total submission to the state. We have read into passages like Romans 13 the Constantinian view of the state. According to this pattern, the church and state are partners. At the same time the teachings of God and Christian ethical standards fade into the background. Violence and evil become rationalized in the "holy war" framework. It is the Constantinian view which breeds the toleration of the CIA's denigration of the value of human life because of exaggerated threats to "national security." It is a warped view of the state which prompts scientists to engage in research which threatens the very future of mankind, just because a government agency has sponsored it or tolerates it.

The Christian's view of the state must begin and end with a realistic view of the state's shortcomings. Even while we strive for justice in our social institutions, we know that the state is not a mirror of divine perfection. The United States is not a Christian nation, in spite of our early heritage that was influenced by spiritual values. Let us not forget to read the passages in Revelation showing power at its worst. There is clearly a basis in Scripture for selective disobedience of the state, as long as one recognized his accountability for his actions. The stance of the Christian may well involve a mix of submission and cooperation with abstention and resistance. As John Howard Yoder has put it: "No state can be so low on the scale of relative justice that the duty of the Christian is no longer to be subject; no state can rise so high on the scale that the Christians are not called to some sort of suffering because of their refusal to agree with its self-glorification and the resulting injustice."[4]

THE STATE AND HUMAN ENGINEERING

These fundamental considerations provide us with a base for examining the specific activities of the state toward human engineering. Reflecting a spirit of national competitiveness in science, as well as a genuine striving for truth and excellence, public support for scientific research has been greatly expanded in post-war America. Considering

240

research alone, federal support has grown from $87 million
in 1940 to $4.1 billion in 1974. A substantial portion of
this publicly-funded research involves human subjects in
"human engineering" experimentation. Recent WASHINGTON POST
headlines about behavior modification noted that some $3
million is being expended for research and application of
behavior modification, involving some 10,000 patients of
psychotherapists and 30,000 to 50,000 residents of institutions.

Those who keep in touch with public funding of scientific
efforts know that the process of authorizing and appropriating
funds for the National Science Foundation has exposed some
nerves which run deep into the painful feelings of the con-
stituents. You have read the news stories about questionable
projects which have prompted citizens to protest to members
of Congress. Some citizens simply do not understand and
accept the validity of basic research. But others legitimately question the use of tax money to study German cock-
roaches, comic books, frisbees and kids who fall off tricycles.
There may well be some justifiable reasons for these studies,
but I predict that a taxpayer revolt is going to force the
government into sending researchers to private sources for
the support of some of these efforts.

As you well know, federal funding and federal regulation
are inseparable companions. Compliance with the rules can
be costly and unpleasant, but some important and necessary
social gains have come from the power of the bureaucratic
purse. The first government guidelines to protect medical
research subjects were issued in 1953 by the Clinical
Research Center, under the National Institutes of Health.
These applied only to internal research, though. Not until
the early 1960's did we have guidelines for extramural
projects. Similar regulations were developed for the Public
Health Service in 1965 and subsequently for NIH, the Food
and Drug Administration, and for the Department of Health,
Education and Welfare.

There has been a progressive refinement of the concepts
of individual rights, informed consent and risk-benefit
ratios.[5] HEW regulations published on May 30, 1974 and
August 23, 1974 require review boards within institutions
receiving grants to apply federal ethical guidelines to
project plans and to monitor the projects in operation.
The new regulations provide protection for special populations for whom informed consent may be problematical,
including unborn, children, prisoners, and mental patients.

With some exceptions, Congress has not been eager to take the initiative in developing statutes or guidelines with respect to biomedical ethics. Few members of Congress and their staff members have the interest and expertise to develop legislation in these areas unless prompted by events and public pressure. The Senate Subcommittee on Antitrust and Monopoly began hearings on the drug industry in 1959, but the question of safety was not of primary concern until the discovery in 1961 of a number of birth defects caused by Thalidomide. From this problem came the motivation for the Drug Amendments of 1962, introducing much tighter testing procedures before commercial release. The concept of "informed consent" was incorporated into the requirements at that point to protect those testing new drugs.

In the late 1960's, issues came to light which began to stimulate legislative involvement of a broader scope. Scientists began calling attention to questionable practices in research, such as the injection of hepatitus virus in mentally retarded children. Meanwhile, members of Congress and other lay persons began to take note of the possible risks involved in genetic research. James Watson testified on this theme before the House Committee on Science and Astronautics in 1971. The year following, that committee published a substantial report on genetic engineering prepared by the Congressional Research Service.[6] Congressional interest in yet another type of human engineering drew the staff of the Senate Subcommittee on Constitutional Rights into a three-year study of behavior modification projects which were being funded by the Law Enforcement Assistance Administration and HEW.[7] Neither of these committee efforts has resulted directly in legislation as of yet.[a]

The most significant Congressional action bearing on human engineering has been the enactment of the National Research Act of 1975 (P.L. 93-348), Title II of which established the National Commission for the Protection of Human Subjects of Biomedical and Behavioral Research. Hearings along these lines began in 1968, to consider Senator Mondale's proposal for a National Commission on Health Science and Society. His Resolution passed the Senate in 1971, but the House did not act accordingly.

---

[a]Since 1975 bills have been introduced in both the Senate and House and hearings have been held, but no further action has been taken.

The National Research Act adopted a two-stage approach to creating an entity to deal with the ethical issues of bio-medical and behavioral research. According to the statute, the eleven-member Commission is to be dissolved in December, 1976, to be immediately replaced with a permanent National Advisory Council for the Protection of Human Subjects of Biomedical and Behavioral Research. It might be said that the Commission will plan the work and the Council will work the plan. The Commission is in the very early stages of its work. Among its current priorities are psychosurgery, institutional review boards, and the treatment of institu-tionalized patients and inmates.[b]

In addition to the establishment of research guidelines and moratoria there have been voluntary efforts with the academic community to regulate genetic research.[c] As a

---

[b]Since the Wheaton conference, the Commission for the Protection of Human Subjects has been active in fulfilling its statutory mandate, but has not been replaced with a National Advisory Council as provided in the National Research Act. Rather, Congress extended the Commission for one year in the Emergency Medical Services Amendments of 1976 (P.L. 94-573).
Shortly after the Wheaton Conference, DHEW published the Commission's fetal research guidelines and simultane-ously removed the ban on such research. The Department has subsequently proposed changes in the regulations on fetuses, pregnant women and *in vitro* fertilization. (FEDERAL REGISTER, Jan. 13, 1977). These changes have in turn been adopted by the Office of Protection from Research Risks, a division of the National Institutes of Health.
The Commission has also completed reports on prison research, and psychosurgery and the disclosure of research information. In the future, Congress may enact some of the Commission's recommendations, as proposed in numerous bills introduced. Congress will also be deliberating upon the future of the Commission - whether to extend its life, allow it to be dissolved or to extend it with additional powers.

[c]While the initial effort was voluntary, the National Institutes of Health assumed the task of formalizing and modifying the guidelines from the Asilomar Conference. These guidelines (FEDERAL REGISTER, July 7, 1976) now apply to all federally-funded research. Issues still under discussion include patent policies for new developments by the private sector. Numerous conferences have been held in the academic and business communities. Local governments have also struggled with their role in regulating research.

243

result of the concerns about the research, a committee of the National Academy of Science invoked a temporary ban on work with recombinant DNA, regarded as being particularly hazardous and unpredictable. The conference at Asilomar this year produced some recommendations for making the research reasonably safe. Those who assume that only government can provide effective regulation of research should note the relative success of this effort.

THE CHRISTIAN COMMUNITY AND HUMAN ENGINEERING ISSUES

In conclusion, I want to make some suggestions related specifically to Christian scholars and professionals. First of all, let us avoid the pitfalls of the Scopes Trial by broadening this debate. It seems to me that this would be one area to demonstrate Christian brotherhood by avoiding evangelical cliquishness and cooperating fully with Protestants of all theological persuasions, with Catholics, and with all of those who would accept some kind of humanitarian ethics.

In this regard, I speak partly from disappointment at the fragmentation of the "pro-life," anti-abortion movement. Surely it is possible to agree on some principles and objectives. As soon as a new organization is formed, it develops its own distinctives, partly to justify its existence. Such is the history of Protestant denominations.

Another suggestion: Let us practice being the salt of which Jesus speaks. Work within the "secular" professional organizations discussing these ethical questions. Christians as individuals can have a significant impact, in my opinion, through their input to such groups, rather than by spending all their time formulating the "Christian" position with the Christian fellowship. A direct result of this effort could be the placement of Christians on the Commission for the Protection of Human Subjects, its successor Advisory Council, the university review boards and any other entities involved with official policy on human engineering. Being realistic this is probably the only way of having input to these bodies. They may pay little attention to a statement from this conference, they may not read our books, they may not appoint our nominees to their bodies, but they may well be influenced by evangelicals who are active in appropriate professional and scholarly pursuits.

A final practical suggestion concerns your input to Congress. Members of Congress and their staffs may scarcely notice an official statement by evangelicals, but they will

not ignore the information and concerns you present personally to your own representatives. Evangelicals have so
often been ineffective in presenting issues to Congress.
They write about the Bible-reading in space craft long
after the matter has been settled, and they write about
Christian radio stations which may be terminated when this
is only a remote possibility. Evangelicals need to take
note of future issues involving ethics, then go beyond that
into other policy questions dealing with science. Our task
must include bringing our ethical and moral concerns into
relevant and prophetic dialogue with those holding policy-
making resposibilities.

Professor Joshua Lederberg, Nobel prize winner from
Stanford, who has been at the heart of the controversy over
DNA research, was recently quoted as saying: "Two years
ago I was predicting that in ten years we would be where we
are today (in genetic research)."[9] Evangelicals must work
very hard in applying ethics to science so it does not take
us ten years to do what needs to be done in two years.

FOOTNOTES

[1]Robert D. Linder, "Fifty Years After Scopes: Lessons
to Learn, A Heritage to Learn, A Heritage to Reclaim,"
Christianity Today, (July 18, 1975), pp. 7-10.

[2]Frederic Wertham, *A Sign for Cain: An Exploration of
Human Violence* (London: Robert Hale Ltd., 1966).

[3]*Washington Post,* (July 11, 1975).

[4]John Howard Yoder, *The Christian Witness to the State*
(Newton, Kansas: Faith and Life Press, 1964).

[5]*Federal Regulation of Human Experimentation, 1975.*
Subcommittee on Health, Committee on Labor and Public
Welfare, United States Senate, (May, 1975). U.S. Government
Printing Office (48-2730).

[6]*Genetic Engineering: Evolution of a Technological Issue.*
Subcommittee on Science, Research and Development, House
Committee Printing Office (85-714).

[7]*Individual Rights and the Federal Role in Behavior
Modification.* Subcommittee on Constitutional Rights,
Senate Committee on the Judiciary, (November, 1974). U.S.

Government Printing Office (38-744).

[8] Alan Fitzgibbon, "Fetal Research Funding Resumption Urged," *Medical Journal*, (May 28, 1975).

[9] *Washington Post*, (July 1, 1975).

EVANGELICALS AND HUMAN ENGINEERING POLICY

John Scanzoni
Indiana University

Senator Hatfield addresses himself to at least five important issues that are fundamental to an understanding of human engineering and man's future from a Christian perspective.

ARROGANCE

First, Hatfield would strike down any arrogance we might hold regarding our recent interest in the protection of human subjects. At times the impression has been given that unless evangelicals begin to charge into the breech, little or nothing of value will be done to protect persons made in God's image from the evil ravages of experimenters. According to Hatfield, much has been done for many years by persons in and out of government who are not avowedly evangelical. Evangelicals ought to be humbled that over the years we have done so little in this area. Moreover it seems apparent that even if evangelicals continued to do nothing (for example, if this conference had not been held), organizations and interested persons in the larger society would at the very least continue to do as much as they are doing right now. Therefore, at the outset we need to dismiss any arrogance or smugness we might possess. Instead, we must humbly ask how we can join with the efforts of those who have preceded us for many years.

JOHN SCANZONI is a graduate of Wheaton College and holds the Ph.D. in Sociology from the University of Oregon. He is currently Professor of Sociology at Indiana University. His main research interests center around changing family patterns and include studies of both whites and blacks with regard to men's and women's work behaviors, fertility control, gender roles and decision-making. Dr. Scanzoni has published numerous articles in professional journals, and has written several books including: THE BLACK FAMILY IN MODERN SOCIETY; SEXUAL BARGAINING; SEX ROLES, LIFE-STYLES AND CHILDBEARING; and MEN, WOMEN AND CHANGE (with Letha Scanzoni).

247

In the past evangelicals have been guilty of undue fear
and suspicion towards researchers and scientists in general,
and especially towards those who have been doing the kinds
of manipulations being examined by this conference. We
have also been suspicious of nonevangelicals who have sought
to bring this manipulation under scrutiny. For instance,
Hatfield asks why evangelicals have not cooperated more
fully with concerned Protestants of all theological persuasions, and with concerned Catholics or Jews - or indeed with
any concerned humanist. Our fear and suspicion may be based
on the implicit notion that if persons or groups are not
evangelicals, then their efforts at establishing ethics will
be below that which we could do. We have been reminded that
we believe humankind to be made in the image of God. Some
evangelicals give the impression that because of the special
or sacred aura we attach to life and to human beings we will
be more diligent than others in protecting God's image.

It can readily be seen how fear and suspicion relate to
the arrogance discussed earlier. If persons do not hold
the exact theological views we do regarding God's image and
other theological issues as well, we label those views less
than adequate. Concomitantly, we fear their efforts at human manipulation will also be subpar, and that we alone can
and must rush in to stem the tide.

Fear and suspicion of researchers solely because they are
not evangelicals is, of course, baseless. Being an evangelical is no guarantee that one is going to be concerned for
the protection of human subjects, any more than being a
Christian led the slaveowner to release his slaves or even
to treat them with dignity, for example through manumission
(allowing slaves to earn their freedom). Likewise, Bible-
believing evangelicals have often been some of the strongest
segregationists. Evangelicals have fought in armies and
killed other people made in the image of God. Some of the
German scientists who were so free with human subjects in
the 30's and 40's may very well have been as orthodox as
we are.

It therefore follows that we should dismiss, along with
our arrogance, fear and suspicion of researchers or other
interested parties solely because of their theology. Instead,
we should be aware that any researcher, Christian or not, is
tempted to violate the rights of human subjects. We should
be realistic enough to recognize that temptation can happen
to anyone at anytime. We must therefore build in safeguards

to hold researchers accountable for the power we give them. The point is that the maxim on which all interested parties, regardless of theology, can agree and therefore act is, what is right, fair, and just to the subject person; what is best for his/her interests?

VIEW OF AUTHORITY

A third challenge posed by Senator Hatfield pertains to the static view of authority held by too many evangelicals. We tend to be rigid conformists when it comes to dictates of both the state and scientific researchers. Such attitudes are paradoxical given the fear and suspicion towards scientists just described, but evidently Christians can be inordinately suspicious of authority and still do nothing meaningful to resist it in the fashion described by Hatfield (e.g., "selective disobedience of the state"). As Hatfield shows, it was precisely that mentality which caused most German Christians to stand by while the law allowed researchers and politicians to exploit innocent victims. Furthermore, Hatfield suggests that mentality has had a role in the silence of American Christians towards recent research in chemical and biological warfare. Under the facade of "national security" we have, much like the Germans, unquestioningly accepted what our government has done. Hopefully there are only a minority of Christians who would identify with the T.V. preacher who actually said, "I love America - yes sir, God has great things for her if she repents - yes sir, I'm a Bible-believer and a redneck and proud of it. You want to know where I am? I'm to the right of you - wherever you are I'm to the right of you!"

Unfortunately our interpretation of Romans 13 and similar passages has led some of us to confuse stagnation with order. We have assumed that to resist authority is to invite chaos and anarchy. Precisely the opposite is true. To fail to resist and change unjust authority is to invite violent and convulsive upheaval - witness virtually any revolution, including our own. The situations of American and South African blacks present similar illustrations of violence emerging when injustice is not resisted by whites of "good will."

In short, Senator Hatfield calls us to a healthy skepticism (not cynicism) of the intentions of both politicians and scientists when it comes to any laws or research that affect human lives through the sorts of manipulations before us this week. Once more we must reiterate the theme that scientists and politicians are accountable for any power we allow them to exercise. Moreover, some evangelicals may

want to consider joining with those who call for the ter-
mination of government-sponsored work whose sole objective
is to destroy (e.g., "germ warfare"). Whatever the issue,
the record of history shows we don't dare to neglect using
whatever peaceful means at our disposal, including civil
disobedience, to achieve justice in the protection of human
beings who happen to be subjects.

EVANGELICAL CONTRIBUTIONS

A fourth issue implicit in Hatfield's discussion is
whether or not evangelicals have anything unique to contribute in the way of concrete public policy suggestions
regarding research ethics. Specifically, would an evangelical who manipulates human subjects behave in any way that is
different (i.e., more ethical or upright) than a nonevangel-
ical? Earlier we indicated that, because of the inherent
selfishness of all human beings, evangelicals require safeguards as much as anyone else. Now we are asking if there
is a standard of "higher" scientific ethics to which evangelicals call others because of our theology. This question
has been raised and debated both in public and private at
many points throughout the conference. There seems almost an
uneasiness to face up to it, because if we reply in the af-
firmative we find ourselves unable to spell out what these
unique ethics are. But somehow to respond in the negative
seems disquieting – as if evangelicals ought to be able to be
uniquely prophetic in this realm.

What is called for is simple honesty. We evidently do not
have anything unique to contribute in terms of calling scientists and politicians to the highest ethics attainable in the
protection of human subjects. However, we can certainly draw
comfort from the argument that over the centuries Christian
notions of justice have so permeated western society that today the high-minded humanist is committed to exactly the same
concern for equity that moves the Christian. Our "leaven"
and "salt" have worked so well that uniqueness in this realm
is virtually impossible to sort out. Nevertheless we can and
should join with humane persons of all persuasions in calling
vigorously for proper safeguards based on the ethics that have
their basic roots in the Judeo-Christian heritage.

In recent decades, as Senator Hatfield notes, evangelicals
have developed the reputation not only of being indifferent
to scientific ethics, but also of being against scientific
progress. Perhaps the greatest outcome of this conference
is that evangelicals have a chance to reshape our image on
both counts. Not that the image *per se* is important by itself,
but as it helps to remove stumbling blocks to personal
250

commitment to Jesus Christ. No one may say after this conference that to be evangelical is to fail to call people to
high scientific ethics, or to be anti-science, or to hold
unrealistic views of unique ethics which do not exist. If
any of these three elements kept one back from faith in
Christ, that no longer should be the case.

Having said all that in no way undercuts the unique motivation that moves Christians who are scientists to do their
work for the glory of God.[1] Though the outcomes in behavioral ethics might not be visibly different from those of
non-Christians, the process of doing it is different in terms
of a calling from Jesus Christ. This sense of mission, vocation, of holy work or service, is an added rationale for
the Christian to do his research as well as he possibly can.
It spurs him to behave in the most just and equitable fashion possible towards human subjects. It is a process or
relationship based on faith that Christ has called him/her
to this work and rewards him/her for being faithful in its
execution. This relationship is indeed a highly rewarding,
ongoing, and growing experience because it places one's
research in a larger and personal context of meaning - one's
work is done in partnership with God. For Christians with
such faith, this relationship is undoubtedly the basic and
fundamental reason to strive for and subscribe to the highest
ethics attainable in scientific research.

INPUT STRATEGIES

Finally, Senator Hatfield's address calls us to develop
specific strategies to implement our inputs into the larger
scientific and political communities. One wishes that the
Senator had gone into greater detail on concrete means to
exercise political influence on Congress. In regard to his
excellent suggestion that Christians ought to seek to exercise greater influence through professional organizations,
we are reminded that in the past we have not overtly encouraged our young persons to seek to serve God through either
applied or basic research. Comparatively few evangelicals
carry on active research programs within the natural and behavioral sciences. Evangelicals maintain a hierarchy of
"eternally significant" vocations; we still use the anach-
ronism, "fulltime Christian service." At the top of the
hierarchy are the clergy - first foreign, then domestic.
Teachers rank somewhere below that. If an evangelical somehow gets an advanced degree in biology or psychology, for
example, we encourage him/her to prove he/she are not suspect. Most often we try to point him/her to a Christian
college where, unfortunately, the teaching and administrative

251

load is so great that seldom can serious, frontier research of the sort we are considering be undertaken.

We have rarely challenged young men and women to examine their gifts to see if they can glorify God through extensive research efforts. It is plain that if there were more Christians among the distinguished leaders in frontier research within the several disciplines where human subjects are manipulated, we could be more certain of evangelical representation and inputs when decisions and laws about ethics are being made. As previously discussed, our own inputs would not necessarily be unique or "higher," but we would be in the councils of power with some sense of participation in the decision-making. One way to begin to alleviate evangelical feelings of powerlessness is actively to encourage our youth to seek out the "holy work" of research. It needs to be emphasized that if God calls them to it, it can be just as eternally significant as translating the Bible in Biafra or preaching in Peoria.

Hatfield has delivered us from the vague generalities and pious platitudes that all to often afflict evangelicals. He has challenged us in at least five different ways to think very hard about the practical implications of Christian concern for the protection of human subjects. If we take these challenges seriously and begin to work towards implementing some actual strategies derived from them, we shall go a long way toward fulfilling the objectives of this conference.

FOOTNOTES

[1]John Scanzoni, "The Christian View of Work," in C.F.H. Henry, ed., *Quest for Reality* (Downers Grove, Ill.: Inter-Varsity Press, 1973).

252

THE PROPER FUNCTIONS OF THE STATE

John A. Olthuis
The Committee for Justice and Liberty Foundation

Senator Hatfield's continuing leadership in relating
biblical principles to complex political problems encourages many people, including Canadians, struggling to do the
same. Accordingly, I am delighted to respond to his paper
and will relate my comments to the following matters:
1) The biblical idea of the nature and task of the state;
2) The idea of the state presently operative in North
America, and the challenges such poses for those Christian
scientists and politicians concerned about the development
and implementation of bio-medical possibilities.

THE BIBLICAL IDEA OF NATURE AND TASK OF THE STATE

Senator Hatfield asked the question "What is the Christian view of the state?" and answered with statements such
as:

I see in the Bible a clear portrayal of the
state as an imperfect but necessary means of
achieving order in society...In a fallen world
awaiting its final redemption, the 'powers'
are ultimately used by a sovereign God for His
purposes. Yet the Christian has been set free
from their dominion and authority through the
Lordship of the resurrected Christ. The state,
then, has a provisional and temporary role, but
a legitimate and important one in maintaining
coherence in human society.

JOHN A. OLTHUIS, B.A., LLB, Legal Counsel and Research
and Policy Director, Committee for Justice and Liberty
Foundation (CJU). CJU is an independent Canadian people's
movement which seeks to develop political, economic and
social policies and programs from a Christian life-perspec-
tive. Mr. Olthuis is author of A PLACE TO STAND and co-
author of MORATORIUM.

I believe his comments express a tension that has long plagued a Christian approach to politics. That tension is the failure to clearly distinguish between the normativity of God's good creation, which remains after the Fall into sin, and man's sinful response that is normative in defining the task of the state. If we see the state as an "imperfect but necessary means of achieving order in society" we lose a normative base for defining "order" and "the state," and for judging how well or poorly government is promoting order. Government is not imperfect because the Fall into sin doomed the state to imperfection. Rather, it is imperfect because of the inability of redeemed, but still sinful, men to respond in perfect obedience to the task God assigned the state.

In his address, Dr. MacKay said that man's task is to glorify God and that man strives to do so at three levels of relationships: individual, family and corporate. He went on to define human "improvability" in terms of the enhancement of man's capacities to serve God and fellow man in his various relationships. I believe that the state's task is to provide and maintain a dynamic public framework which encourages and enhances the diverse societal possibilities man has to grow in praise of God and service of fellow man. Bernard Zylstra lucidly discussed that task in a recent article. The following is a brief summary of his views.[1]

Zylstra says that Christ's redeeming compassion over the hungry multitude, the poor, the suffering ones, must be also expressed by Christ's body because we cannot understand the meaning of Christianity if love of God is separated from love of our fellow man, and vice-versa (cf. Deut. 6:5 and Lev. 19:18). The New Testament throughout posits the inseparable conjunction of the two loves (Matt. 22:36f.). "Love does no wrong to a neighbor; therefore love is the fulfilling of the law," (Romans 13:10). "He who does not love does not know God" (I John 4:7).

The two love commandments sum up whatever God requires of human beings. Justice and stewardship are not commandments from God that stand next to the commandment of love. Rather, they are specific instances of the ways in which we are to love our fellow man. Although it is impossible to capture exhaustively the content of justice in human language, Emil Brunner's definition is particularly helpful. Brunner says:[2]

The Christian concept of justice...is determined by the conception of God's order of creation. What corresponds to the Creator's ordinance is just - to

254

that ordinance which bestows on every creature
with its being, the law of its being and its
relationship to other creatures. The 'primal
order' to which everyone refers in using the
words 'just' or 'unjust,' the due which is rendered to each man, is the order of creation,
which is the will of the Creator made manifest.

Zylstra arrives at this provisional summation:

> ...the norm of justice requires a social order
> in which men can express themselves as God's
> imagers. To put it in different words; the norm
> of justice requires social space for human per-
> sonality. By personality I then mean the human
> self whose calling lies in the love of God and
> love of fellow man. That calling entails the
> realization of a multiplicity of tasks in history.
> Justice therefore also requires societal space
> for man's cultural tasks. Moreover, the reali-
> zation of man's central calling also entails the
> establishment of social institutions, like mar-
> riage, the family, schools, industries and the
> like. Hence, justice requires societal space
> for these institutions as long as they contrib-
> ute to meaningful, harmonious and an opened up
> human existence.

In discussing Romans 13, Zylstra says that "Authority is
office; that is, a channel for the realization of divine
norms in a social relationship." Accordingly the person in
authority is "God's servant for your good." (Romans 13:3f.).
The state and its authorities, Zylstra says, exists for the
good of the citizenry. This he says in a nutshell, is the
gospel's message for politics. Politicians are office bearers. They are to execute their legislative, judicial, administrative or executive offices only for the good of the
citizenry. That good is public justice. The state, then
must provide the framework within which men have full freedom to choose whether or not they wish to glorify God.

In this context, I think it important that Senator Hatfield clarify what he means when he says:

> The Christian's view of the state must begin and
> end with a realistic view of its shortcomings.
> Even while we strive for justice in our social
> institutions, we know that the state is not a
> Christian nation, in spite of our early heritage

that was influenced by spiritual values.

Is he saying that the United States is not a Christian nation because the government does not accept the biblical view of man and the state or because, although accepting the biblical view, its imperfections are such that it has lost its Chrisitan character?

This distinction is important because every person and institution acts on the basis of some view of man and society. If it is not a biblical view it is important to discover what view it is because of the critical implications for policy decisions on all matters, including human engineering.

THE IDEA OF THE STATE PRESENTLY OPERATIVE IN NORTH AMERICA

My conclusion is that North American governments do not embrace a biblical view of man and the state. The dominant and culturally formative view is that man is essentially an economic being. His happiness varies in direct proportion to his accummulation of material wealth. The task of the state, therefore, is to set the public conditions which enhances that possibility. Science-based technology, with its capacity to increase per capita productivity, has come to be viewed as the state-encouraged vehicle to deliver happiness.

The religious pursuit of technology is encouraged by an economic and political framework which favors economic growth above all other objectives. In fact, the framework which favors it to such an extent that it severely limits the possibilities for the proper and harmonious development of the many non-economic sides of life. Let me mention some of the widespread and serious human distresses that I attribute to this reductionistic one-dimensional view of man. Pollution and resource depletion are well-known and obvious results. Let me suggest a few others:

1) The alarmingly high incidence of alcoholism, drug addiction, and suicide, added to the rapid breakdown of interpersonal relationships like friendship and marriage, combined with more emotional illness, sexual infidelity and street and political crimes, are ever present symptoms of that distress.

2) The Western world's exploitation of Third World resources seriously interferes with the possibilities Third World peoples have of using the share of the world's natural resources intended for them. Starvation, malnutrition, disease and

economic and cultural poverty are some of the more obvious results.

No one consciously aimed for these results. In fact they are generally viewed as the undesirable but necessary trade-offs necessary to achieve the aggregate increases in Gross National Product required in order for the material benefits to trickle down to every member of society. In theory, once society achieved quantity, it could turn its attention to the quality of human life.

But, reality is showing that attempts to concentrate on quantity before quality are inherently self-defeating. The position that the quantity of material possessions is the key to man's happiness is a religious commitment. It leads the faithful to believe that happiness, although not yet achieved, must be just one more technologically-induced increase in GNP, or just one more car around the corner. At some future time we may well be told that happiness is one more clone or one more psycho-surgical modification of aggressive behavior around the corner!

Admitting that science-based technology is not the way to happiness requires the confession that our gods have failed us, a confession as abhorrent to doctrinaire believers in science as a denial of the virgin birth is to the fundamentalist Christians. But growing numbers of people in our society are making that confession. It pains me to see evangelical Christians clamouring to board the science and technology bandwagon just when so many non-Christians are jumping off and looking for new values by which to guide their lives.

These new values must admit the self-defeating nature of the technologically-induced quantity-before-quality designation of human needs. They must advance the belief that food, clothing , shelter and like needs are but some of the many human needs that must be met in a simultaneous and integrated, rather than a hierarchical-quantity fashion. The quantity of goods, and accordingly, the use of technology, must be determined and not be determinative of the needs of an integrated quality lifestyle.

I am suggesting that science-based technology must bear a considerable burden of blame for contributing to the very problems it now claims it is developing the capacity to solve via human engineering. The growing awareness that nature is finite could have led man to more easily accept that he too is finite. Rather, we seem to be headed in the direction of

trying to make up for nature's finiteness by attempting to
make man more infinite. This non-biblical view of man and
the state means that most discussions about public policy
and human engineering possibilities end up viewing the judicial and legislative branches of government as facilita-
tors. Their task is to educate and prepare the public for
the staged introduction of scientific possibilities.

Senator Hatfield's description of the attitude and relationship of the United States Federal Government to developments in human engineering indicates that the United
States government is a facilitator. Recognizing this, it
is, I believe, urgent that Christians respond to the Senator's call that they become involved in the scientific and
political decision making processes. This is particularly
important because of the "drift" inherent in the facilitating position. I was particularly struck by this when reading an article by J.G. Castel on the "Legal Implications of
Bio-medical Science and Technology in the Twenty-First Century."[3] Castel says,

> I predict that, within the next fifty years,
> legislation will be concerned with removing any
> existing barriers to the full application of new
> bio-medical discoveries and techniques subject
> to reasonable and necessary controls...The tech-
> niques such as organ transplants, surgery *in
> utero*, ovum and embryo implantations, clonal re-
> production, cryonic suspension and behavior con-
> trol. Thus our society must make sure that exist-
> ing forms of political decision-making are adequate
> to the task of using the powers given by geneticists
> to their best advantage...For the first time in
> history man has the potential to improve the
> conditions of the entire human race, or to destroy
> it; let the legal profession not miss this ap-
> pointment with destiny.

In discussing therapeutic insemination of ovum, or embryo
implantations and uterine transplants, he says these "will
be legalized even in the case of unmarried women." "The
law" he continues "should no longer consider the family as
a biological or genetic unit but a consensual and spiritual
unit." Castel makes it clear that bio-medical possibilities
should determine the nature of marriage and the family. He
indicates that the law should facilitate such by enacting
whatever laws are necessary to implement the transition. I
think a more biblically sensitive view is that the law should
facilitate what encourages a biblical view of marriage. It

258

should discourage what interferes with such. In this
context I agree with Dr. MacKay's statement that,

Health is good. But artificial stock breeding
or cloning the finest specimens of manhood at
the cost of destroying the biblical ideal of
family relationships would buy health at too
great a cost.

Castel goes on to say:

Once behavior modification to modify aggressive
actions and to restrain destructive behavior
or antisocial behavior is undertaken (a move,
which Castel supports wholeheartedly, because
it will no doubt improve the health and happiness
of people) in individual cases, it is not at
all certain that our society will resist widespread resort to such manipulative techniques.

In this context Harvey Wheeler says:[4]

Manipulation is a grave problem in every industrial society: It is an even more alarming threat
to postindustrial scientific man, for science
consists essentially of the theoretical implications of the manipulation of things...This turn
means that as science invades each new realm it
leaves a stream of manipulated things in its wake
...Take the example of the difference between the
traditional profession of medicine and modern
medical science. Traditional medicine was one of
the humanistic professions. Its ethic proclaimed
that patients were invisible and were to be treated
as ends in themselves rather than means. Medical
science, like all science, deals only with things,
and this converts patients into objects of research,
experimentation and manipulation. This is the
setting in which we must consider the new manipulative devices now appearing in many different
forms. (He then mentions devices like electrodes
planted in the brain, and mind bending behavioral
drugs and continues...) In short we now have, or
soon will have a battery of manipulative devices,
that were employed by Hitler and Stalin. Ellul
states that anything that can be invented will
be used. The history of manipulative instruments
from the first ice age tools to "The Selling of
the President" indicated he is right. Our only

options are to face this cybernetic age and
try to control it or ignore it and let it
control us. We must join together in creating
a new age of wisdom.[5]

Wheeler puts his hope in a possibility based on developments in brain technology. Until very recently, he says,
psychiatrists, brain psysiologists and soical psychologists
were all in agreement about the instinctual-rational division of the brain. This view supposedly explained why
man's violent and irrational animal essence often over-
whelmed his reason. Now, says Wheeler, the new brain
physiologists believe man does not have a Jeckyl and Hyde
brain, but is a rational-emotional creature throughout.
If this is so, Wheeler argues, the very idea of the irrational loses its former relevance and we are on the brink
of a new age of wisdom.

Science and technology promise to join hands to
validate a reborn humanistic philosophy of democracy. It is not that man himself will have
changed, but that his changed view of himself
will profoundly affect his view of the world,
its problems and what can be done about them.

I think Dr. Wheeler's hope is empty. Science and technology will not usher in the new age of wisdom. Neither
will Christians usher in the new age of wisdom. But I
take it to be our task to do our political and scientific
work in such a way that we honor ourselves and our neighbors as God's image bearers. In terms of human engineering, I think this implies that the Christian scientist
need not work in a spiritual vacuum. His particular calling is to develop science and technology in such a manner
that it has the potential to help men serve God. This is
an exciting framework for the development and use of those
bio-medical possibilities that will in fact help man glorify
God.

With respect to public policy, the Christian must work
for a just public framework with gives each man his due
as God's image bearer. This means the introduction of
those bio-medical possibilities which help man in his task,
and the banning of those that tend to detract from his service to God.

If we are willing to accept the human responsibility
this framework requires we need not adopt Harvey Wheeler's
optimism or Jacques Ellul's pessimism. Rather, we can work

out our scientific and political salvation in fear and
trembling, with all the joy and hope of the new Life that
we have in Jesus Christ.

FOOTNOTES

[1]Bernard Zylstra is a senior member in political theory
of the Institute for Christian Studies, Toronto, Canada.
The article referred to appeared in the Fall 1973 issue of
the *International Reformed Bulletin*.

[2]Emil Brunner, *Justice and the Social Order* (New York:
Harper, 1945), p. 83.

[3]J.G. Castel, or Osgoode Hall Law School, York University, Toronto writing in *The Canadian Bar Review*, Vol. Ll,
(1973), p. 119.

[4]Harvey Wheeler is the senior member of the Center for
the Study of Democratic Institutions in Santa Barbara,
California. The quotations are from his book, *The Politics
of Revolution* (Berkeley: Glendessary Press, 1971), pp. 296-
301.

[5]In his integrative summary, Dr. MacKay has dismissed
Wheeler as a "scaremonger." While his warning against irresponsible statements is well taken, I cannot agree that
Wheeler's remarks fall into that category for three reasons:
(1) The manipulative nature of science which Wheeler is
decrying, appears to me to be consistent with the non-
biblical view of man and society which is formative for
the North American Scientific and Political communities;
(2) In his book *The Technological Society* (New York:
Vintage Gooks, 1964), Jacques Ellul (p. 432) mentioned some
of the predictions being made, such as that by the year
2000 natural reproduction would be forbidden and will have
been replaced by introduction into a carrier uterus of an
ovum fertilized *in vitro*, ovum and sperm having been taken
from persons representing the masculine ideal and the fem-
inine respectively. Ellul said it used to be we were hearing these things from science fiction writers. "Now we
have like works from Nobel Prize winners, members of the
Academy of Sciences of Moscow and other scientific notables
whose qualifications are beyond dispute...Serious scientists it must be repeated, are the source of these predictions." Wheeler's comments about the manipulative devices

261

are not his science fiction. They are based on claims
of responsible scientists;
(3) If such claims are in fact exorbitant, is it not high
time that responisble scientists increasingly and consistently publicly say so? Calling a spade a spade if it is a
spade can only help the non-scientific populace in resisting
scaremongers and others of their ilk. Simply dismissing
the challenge with an epithet like scaremongerism is itself
of little help. In fact, today it only weakens the already
strained credibility with which the scientific and technological communities are afflicted.

SCIENCE AND THE NEED FOR VISIBLE VALUE-OPTIONS
IN POLICY-MAKING

Carl F.H. Henry

It seems to me that the deepest concern over technology and public policy in America is whether the swift decisions which a technological era sometimes require are compatible with democratic processes. Is our republican system of democratic government vulnerable in times of crisis, through its reliance on public education and popular choice, to costly delays and inefficiencies that are not tolerated by other governments? The decision to drop the hydrogen bomb on Japanese cities was notably reached by a small circle whose influential participants disagreed. Impressed by contributions of applied science during the Great Depression and World War II, the United States created the National Science Foundation whose budget has leaped from $3.5 million to $700 million in 25 years. Is it likely that in a future international crisis technology could best serve the national interest only by a disregard of democratic values and institutions? Can public policy balance both the interest of technological science and democratic society and preserve the nexus at which each best supports the other?

VISIBLE OPTIONS

Another basic problem in public policy is the identity

CARL F.H. HENRY, theologian and writer, is Lecturer at Large for World Vision International and has served on the faculties of Northern Baptist Theological Seminary (1941-47), Fuller Theological Seminary (1946-56) and visiting professor at Gordon College, Wheaton College, Eastern Baptist Theological Seminary, and the Asian Center for Theological Studies and Mission in Trinity Evangelical Divinity School, South Korea. He was founding editor for CHRISTIANITY TODAY magazine. He is author of 20 books and editor of many others, including BAKER'S DICTIONARY OF CHRISTIAN ETHICS. His most recent publications are a major theological work, GOD, REVELATION AND AUTHORITY and a forthcoming symposium SCIENTIFIC FRONTIERS: CHRISTIAN SCHOLARS SPEAK OUT.

of those who establish the values and goals to be pursued in technological programming. Whether one speaks of a single nation or of the United Nations, the problem of who or what controls the controllers becomes increasingly urgent. Whoever directs and controls the power man has developed may decide the future of contemporary civilization. Science is not itself a source of moral norms. In a democratic society government is in principle answerable to all the people. In respect to public policy issues in technology, Washington is increasingly overrun with advisors. The President has an office devoted to scientific and technological advising. The Senate has technological advisors; the House has them; N.S.F. has them; N.A.S.A. has them. In addition to government agencies there are multiple private advisory agencies. The A.A. A.S. Commission on Science Education alone has issued annual bibliographies containing thousands of titles; the third edition (1972) indexed 86 important books and articles under the Science and Public Policy section. It also referred to other sections on Science and Social Responsibility and on Science and Society. Subsequent editions have supplemented these listings.

Neither Christians nor non-Christians are at liberty to impose their views upon American society by force. In order to promote justice we can and ought to join forces with all groups devoted to the same end. But Christians dare not neglect the persuasive public presentation of their convictions. They need to proclaim the standards by which Christ will judge men and nations. They especially need to speak out in a time when terms like justice accommodate a variety of contents. If Christians do not promote acceptable legal options they will forfeit the formation of public policy to those who sponsor alternative views. On the basis of divine authority Christians hold to fixed values and goals. This is in contrast to those who subscribe only to experimental approaches to life. The Christian aims to persuade the citizenry of the moral rightness of his views. In so doing, even fellow-citizens who personally comprise these standards will nonetheless acknowledge that public law and community conscience should take the high road. Whatever the public's response, Christians must make their public witness. It would be well if Protestants and Catholics could emphasize agreements in the public realm, and not simply disagreements. Disagreements even among evangelicals need not be concealed, but it would be desirable if all Christians would unite on some important issues in a common witness. Christians should also state clearly what alternatives would be so morally objectionable that they could maintain a good conscience only by civil disobedience.

POWER AND THE PUBLIC

We live in the post-Watergate era which is especially
sensitive to the proper and improper uses of political power.
In contemplating the rights of the powerless and deprived,
why shouldn't Congress, the Executive branch and the Supreme
Court establish a commission that conducts hearings to openly
debate the moral and immoral uses of technological power?
Public reaction to experiments in the field of human health
and behavior has already called attention to the need for (1)
balanced professional judgments of probabilities, risks, and
costs; (2) awareness of moral principles and national priorities bearing on the encouragement and support of such experiments; (3) recommendations of peer reviewers (sometimes at
odds, as in heart bypass and electrode implantation surgery)
in view of the present state of evidence, adequacy of available treatment, availability of medical facilities; and (4)
informed consent of participants, assuming that such participation will not endanger public health.

Peer groups all too often reflect professional prejudice,
and frequently lack expertise about moral or legal questions.
Why shouldn't the nation hear from the general public also,
including representatives of the mentally retarded, impris-
oned, and the genetically deformed, and their families? Even
if decisions need not be made on a jury system basis, would
not the conscience of the nation be sensitized, the mind
informed, and the will motivated by the fullest span of information and experience? Public scrutiny is desirable and
necessary, the more so if scientific research is publicly
funded.

COMMITMENT AND CHOICES

The role of the Christian Church as a publicly identifiable
community in American life is rendered doubly significant by
technological developments. The Church of the twice-born is
to serve in the world as light and salt. It is to illuminate
the new society that lives by the standards of the Coming
King, and preserve a putrefying civilization from unbridled
injustice and chaos. The Church can best witness on the national scene when her positions on abortion, human engineering, and other issues are not simply promoted by lobbyists.
Rather, those positions also need to be exhibited as the lifestyle of the fellowship of the redeemed. If secular society
tends to subscribe in technological matters to utilitarian
ethics ("the greatest good for the greatest number"), and even
to the "survival of the fittest," the Church can best show
her concern for the weak and powerless as an evidently

committed community.

There is of course nothing wrong with maximizing the good for the greatest number if one assuredly knows what the good is. But, an outlook that downgrades or dismisses the value of human life if one is elderly or paraplegic or retarded or deformed or an unborn fetus has missed the biblical ethic. Not only does human life have special value in the sight of God, but its voluntary destruction in principle diminishes my own worth. Such destruction also saps human sensitivity to compassion and justice. However, if Christians opt for the preservation of helplessly deformed fetuses, does not social sensitivity also require a supply of financial resources and medical facilities for their proper care? Are the churches ready to be the new community both in doctrine and life? For example, if we protest the availability of kidney machines for the high and wealthy, can we give that protest public force? Are we willing to provide pilot projects which in our own community are available "first come, first served" in reflection of the rain God showers on just and unjust alike?

The existence of modern technology does not give any nation or society a "technological imperative" to pursue all possibilities. Some options are surely more destructive than constructive, and technological science is not infallible in its prognostication of beneficial consequences.

The debate over technology and applied science in the realm of medicine and public health accelerated in 1961 with the discovery of birth defects caused by Thalidomide, which precipitated a demand for more adequate testing. Soon mass genetic screening programs as part of national health policy faced numerous challenges. Although 43 states have approved the screening of newborn babes for metabolism defects that lead to severe mental retardation, early techniques were less accurate than believed, and some babies were placed on harmful diets. In sickle-cell anemia investigation many people confused the anemia trait with the active disease and were frightened; even some insurance companies jumped their rates. In an age when "applied futurism" fascinates some philosophers and politicians it is well to keep in mind that the faulty prevision of technology (e.g., the destructive potential of DDT, belated cancer fears related to the hazards of vinyl chloride in plastic manufacture, unforeseen perilous nuclear radiation from power plants) has given rise to a call for preventive technology.

266

Many persons criticize the tendency to approach public health considerations on a cost-benefit basis. But it would be foolhardy to see only the benefits and to be blind to the cost. Does the federal budget deficit, or national priority in a time of recession and unemployment, or prudence even in normal times, permit unlimited national health commitments? Would the world's financial resources sustain an extension of medical technology to the developing nations which increasingly expect benevolent assistance from the larger powers? At what point do health concerns of the citizenry become and cease to be national priorities? Is it adequate merely to provide national medical information concerning the status of the war on disease and deformity? Should public screening be provided that leaves the burden of decision and expense to individuals? Ought public surgery or institutionalization be provided where victims of severe deformity do not owe their existence to parental neglect of medical knowledge?

EXPERIMENTATION AND ETHICS

The National Research Act of 1974 banned research on any "living" human fetus before or after induced abortion except in order to save the life of the fetus. Some diseases (e.g., the Rh factor) can now be diagnosed or treated because of earlier fetal research. Salk vaccine, the first big breakthrough against polio, came as a result of such research. The fact of some benevolent consequences does not of course establish the moral propriety of any course of action. But if animal studies are first conducted and necessary information cannot be otherwise acquired, why should not nontherapeutic research be permitted with parental approval on fetuses aborted or scheduled to be aborted on moral grounds (e.g., rape; the mother's survival) - particularly if such abortion occurs sufficiently early that the fetus cannot survive (less than 601 grams and 24 weeks)?

Human experimentation of certain kinds threatens the dignity, integrity and individuality of human persons. Not only do "genetic engineering" proposals which would clone multiple copies with identical genotypes collide with these concerns, but other programs raise similar questions of moral legitimacy. One area of growing concern relates to informed voluntary participation in experiments. The WASHINGTON POST reports a $3 million project in behavioral modification involving perhaps 10,000 psychotherapy patients and more than 30,000 institutionalized persons. The U.S. Army has acknowledged, following

the disclosure of facts relating to the suicide death of the
civilian scientist Frank Olsen, that it administered LSD to
nearly 1500 persons in 1956 and 1957. At about the same
time it included some 900 civilians in a testing program in
which some universities cooperated. The purpose of such
tests was to learn whether LSD and other hallucinogenic
drugs are serviceable against enemy forces in wartime. The
use of servicemen or of prisoners for many types of experiment has the advantage of a group situation with a highly
controlled diet and regimen. But without individual consent,
full information and proper compensation or incentives it
would be immoral. The experimental injection of hepatitis
virus in mentally retarded children in the 1960s was rightly
challenged.

PUBLIC INTEREST AND NATIONAL SECURITY

The question of secrecy versus public interest is not
easily resolved. In our century public interest has been
used as a shield for such enormous crimes that one can well
understand the present tendency in democratic societies,
particularly in the U.S., to make everything public. But in
a world with giant predator powers such a policy, however
highly motivated, will in the last analysis be destructive
of national security. It may ultimately contribute to an
undermining of treasured human liberties. To disclose what
would be helpful to a declared enemy is contrary to national
interest. Yet what is done in the name of advanced technology must be done within the law. If the law is held to be
on the side of wrong or injustice, one may work by all legitimate means for its replacements, but the lawbreaker must
stand ready to pay its penalties. Declared or known enemies
and predator powers who place themselves above the laws of
nations imply an invitation to other nations to do the same
thing in the case of self-defense. A whole range of activities from censorship to espionage gains its legitimacy and
necessity from such considerations, but neither justice nor
love provides any moral basis. Totalitarian rulers can decide by one stroke what is in the public interest since the
public interest is seen as equivalent with the interests of
the rulers.

In a democracy some competing claims of public interest
and national security are inevitable. While the conflict
is usually championed as a straight moral issue, elements
of self-interest, sectional-interest or party-interest frequently intrude. Here the Christian espousal of good conscience and just law can be critically important. Neither
national security nor public interest is the absolute concern

268

of an enduring nation, except as both find higher unity in the revealed will of God cherished in private and public life. Servicemen in the hallucinogenic drug program were told not to reveal their participation because of national security regulations. Such experiments are often conducted on the premise that if we have scientific resources for a given project an enemy nation may also, and we should be prepared to cope with any contingency. The argument has its point, of course, and the barbarism of Twentieth Century dictatorships is written too large in the Nazi gas chambers and in THE GULAG ARCHIPELIGO of which Solzhenitsyn writes to shut our eyes. But, instead of being motivated so largely by fear that we do ourselves what is reprehensible, we ought more largely to be motivated to benevolent pilot projects that an enemy would not undertake.

AN ALTERNATIVE

The real challenge to the Christian in the Twentieth Century is not simply to provide lucid criticism. It is to re-create science in an acceptably Christian way. The directions in which man is being modified in the modern era must be of crucial concern to the Christian whose view of man is of central importance and has implications for the currently reigning view of nature and reality. In this respect we should give more attention to the subtle conditioning and compulsive elements in the educational arena, and the implications for informed consent in the realm of learning. Scientific naturalism, or radical secularity, which is now the most formative outlook on the university campuses of the Western world, holds that all existence is reducible to impersonal processes and events. A survey of 400 recently graduating seniors by YALE DAILY NEWS indicates that 54% do not believe in God. It would be illuminating to learn what god the minority believe in.

Contemporary theory espouses the notions that all reality has an expiration date, the space-time sphere is a colossal Las Vegas in which the universe and man are chance products, that all notions of truth and the good are relative to the cultural context in which they arise, that man himself is the time-bound sponsor of whatever future structures will shape nature, history and society, and that man's maturity consists in his deliberate rejection as myth of all assertions of God and of the cosmic ultimacy of personal values and an enduring truth. The reluctance of Western leaders to emphasize the great concerns of conscience, spirit, enduring truth, transcendent justice and the reality of God dipped to a sad nadir in President Ford's cancellation of a

269

meeting with Solzhenitsyn and his disappointingly trivial
satellite comments on the occasion of the Apollo-Soyuz dock-
ing while the world watched and listened.

The Christian revelation unmasks as myth the conjectural
metaphysics of a cadre of influential intellectuals who
arbitrarily transfer a wholly deserved respect for science
to a naturalistic world-life view that has no more basis in
science than would the worship of liverwurst.

PART VIII
CONFERENCE PERSPECTIVES

HUMAN ENGINEERING AND THE
CHRISTIAN CHURCH: SUMMARY AND REFLECTIONS

Donald M. MacKay

How should Christians view human engineering? Seeking
the way of humility, our first reaction might be strongly
negative: "I'm content with what God gives me; I don't
want to interfere." The reaction may be reinforced by
sheer inertia. "It's dangerous. We don't know enough.
Where will it all lead? Best keep out...let the world get
on with it if they will."

But will this do? "He that knoweth to do good and doeth
it not, to him it is sin." It appears from these new developments that the sum of misery in the world is reducible.
God is the Giver of the new knowledge. It is He who will
one day ask: "What did you do with it?"

At the outset, Dr. Callahan raised the key question:
"Do we have a positive obligation to do good, or is our
obligation only to avoid doing harm?" In response it was
generally agreed that the Christian cannot stop at avoiding
harm. We do have an obligation to do good, if the good is
well identified and in our power.

The first thing we are faced with, however, is the ever-
present risk of *superbia, hubris,* human pride. Even if
Christianity rejects in principle all pagan and supersti-
tiously fearful attitudes towards the natural world and
natural laws, self-glorification is a constant temptation.
Dr. Spencer reminded us that neither self-glorification on
the one hand nor terror on the other are appropriate responses to the biblical perspective on our human situation.

Secondly, Dr. Callahan reminded us that our power is
bonded, limited power. It is an illusion to think that we
can proceed without limit in any of these directions, because
sooner or later costs catch up with us. It is therefore
essential that we go slowly and, if possible, reversibly.
We must remember that we are not only finite and limited in
our wisdom, but also sinful, therefore warped in our motives.

273

Third, even good aims can conflict, especially between the different levels — individual, family, corporate — at which human fulfillment is to be sought. For example, reduction of infant mortality, which is surely an individual and family good, conflicts with the aim of preventing mass starvation, unless we can find a humane and acceptable way of avoiding exponential population growth. There are many examples where it is not a simple matter of choosing whether or not to do good. Rather, it is the dilemma of wondering whether we could ever see clearly enough to add up the sum of good and evil, and work anything out as a clear and final answer. We are continually fumbling for an understanding of the controls of an exquisitely complex mechanism, which we can all too easily wreck. We shall need all the wisdom that its Creator can give us if we are not to do more harm than good by our intervention.

Fourth, the achievement of material goals and improvements can all too readily swamp the spiritual point and purpose of our human existence. We remember the rich man in Christ's parable[1]: "Soul, thou has much good laid up for many years; take thine ease, eat, drink, and be merry." And the answer of God, "Thou fool, this night thy soul shall be required of thee." The question with top priority is always: "How will it all end? To what end is it all being directed?" These things can conflict miserably.

Fifth, the manipulative approach, even when well intentioned, can degrade human subjects. We must reject the sweeping generalization that medical science converts people into things as typical of the sort of extremist propaganda which brings discredit upon arguments that might otherwise deserve respect.

Sixth, there are no human engineering substitutes for personal salvation. This is true even if some such as Dr. Clement feel that the virtues listed as "fruits of the Spirit" can be assisted by the kind of reinforcement (or perhaps encouragement is the ordinary word for it) that behavioral psychology is able to offer.

PRINCIPLES AND SITUATIONS

Where do we turn for guidance in such a maze? Will the old Judeo-Christian values still serve? An immediate answer is that values or moral criteria don't serve us at all. They judge us. But it is a good question whether the old slogans will still serve to articulate the relevant biblical criteria as applied to these new situations. Take, for example,

the slogan of "the sanctity of life." This can be confusing
if we take "life" in too strictly a biological sense. We
have responsibilities to God as procreators to do the best
we can with the data He gives us to bring God-glorifying
lives into being. To use only the slogan "the sanctity
of life" to determine whether a fetus should survive for
example, seems to many inadequate and simplistic.

Again, the slogan of the sacredness or worth of the
individual is admirable, and thoroughly biblical as applied
to the normal grown human being. But in borderline cases
we may have to ask whether we in fact have an individual
person here to whom it is meaningful to attribute rights.
We sense here the difficulty of the duty to tread the middle way of biblical realism. This is a narrow path between
on the one hand, an arrogant lack of respect for the fullest
potentialities of the biological situation that exists before a conscious child comes into being, and on the other
hand superstitious and meaningless talk of "responsibilities" to non-persons. We have to recognize that the fetal
situation at an early enough stage is essentially a physical
and biological, not a personal, one whatever the potential-
ity may be. In all this the Creator is beside us, knowing
the facts better than we. He is affronted if we underes-
timate through carelessness or any other unworthy motive
the personal capacities of that biological situation. By
the same token, we must remember that if in God's sight,
in a particular abnormal case, there is not anyone there
with a claim on us, then we will do Him no service by going
through pious or superstitious contortions as if there were.
Nobody would wish to minimize the difficulties in practice
in determining what is in fact the case. However, at least
it should help if we can get straight the questions for
which we need answers.

In this connection it is important to beware of an illicit
and confusing form of argument that I might term "Thin-
end-of-the-wedgery." This (a twin brother of "Nothing-
buttery") often crops up when people ask, "At precisely
what point in time do we have a fully human individual
with rights?" This sounds like a sensible and even an
urgent question: If we cannot justify a precise answer
the "thin-end-of-the-wedger" is liable to argue that there
is then no real difference between a conscious human infant
and a fertilized egg, or between a responsible human agent
and a brain-damaged "human vegetable."

The logical fallacy is exposed if we consider a parallel

case. Nobody can rationally establish an exact number of
hairs, N, such that anyone with N hairs on his chin is
bearded and anyone with N-1 is not. But this in no way
proves that there is no real difference between bearded
and being beardless. In all such cases we recognize the
difference by looking for contrasts between the ends of
the continuous spectrum, not by discovering a precise di-
viding line.

So it is, I think, with the way we should think of the
development of the embryo. The search for a precise point
at which we can prove that we have a "living soul" may be
vain. However, this in no way tends to debunk or reduce
the real distinction between an object that is the body of
a living human person, and an object that is too immature
or too deformed to be so.

The same point arises when we ask under what circumstances
it is meaningful to seek the "informed consent" of a mentally
defective patient before operating. The suggestion was made
that when either immaturity or infirmity made true dialogue
impossible, the ethics of proposed treatment might still be
checked by considering what answer one would make to an
imaginary advocate. For the Christian, Christ Himself is
always a real advocate in that capacity. We must each
answer to Him in sober truth at the bar of judgment. When
the fullest attention to the available facts, including the
data of Scripture, leaves us perplexed, it is in dialogue
with Him that the Christian has his most realistic resource
for the good of his patient and those he seeks to serve.
We must ask His Spirit to illuminate for us the relevance
of His revealed will and the other data we have. No casuistic
book of rules, however expedient in our sinful world, offers
an adequate subsitute for this experimental test that the
Christian servant can and must make.

It is important however, that we should distinguish
between this insistence on the need for direct reference
to Christ for the wisdom of His Spirit, and what is popularly called situation ethics. The point is not that in
these cases a single clear biblical law applies. The point
is that we are confronting situations where several biblical
principles (respect for human life; compassion for other
people including relatives; desire that God may be glorified
by the fulfillment of human possibilities, and so forth)
seem to tug us in different directions. This is the sort
of situation where I believe reliance on the Holy Spirit
to show us the relevance of Scripture, and to illuminate
our minds to see the relevance of other important information

276

is meant to be a reality for us. This is something very different from thumbing up a rule book. In the same way we must be careful to distinguish between what one speaker referred to as the continual transformation of the Christian mind, which we recognized as a Christian duty, and what is popularly advocated as the revision of our values in the light of new knowledge. Someone quoted C.S. Lewis as remarking that you could no more expect to discover new values than to discover new primary colors. The kind of openness that we recognized as a Christian duty can never be expressed by way of blindness or disobedience to revealed truth.

So far I have been summarizing points of caution. The Bible also has much to say on the positive side. Not surprisingly, very little of this is in the way of direct commandment. Encouragement comes more indirectly from the biblical perspective and biblical priorities.

First among these, for the scientist and the human engineer himself, is the most general principle of all: "Seek ye first the kingdom of God and His righteousness" (Matt. 6:33). This we found to be a very thorough-going one, especially bearing in mind all the risks of counter-attractions.

Second, God's first priority for the people we are seeking to serve, is that they should be enabled to glorify and enjoy him forever. To that end, the Bible urges upon us the creation ordinances of marriage and family life, and the moral ordinances of the Law. Particularly relevant are the values of fidelity, integrity, loyality, obedience in the family and in corporate relationships agreeable to the law of God.

Are these truisms? They are certainly familiar enough; but as I have already suggested, to work through what these things should mean in particular cases is necessary if we are to understand God's will in each case. What does it mean in this context, for example, that man is made in the image of God? Primarily, no doubt, it means that he is answerable to God. Man can be 'Thou' to God, and knows what it means to be challenged by God. It also means that we are meant to be like God. In particular, God is dead straight, so our being in the image of God means that we are to be dead straight. I feel this is a major key to our problem. Almost every procedure we have considered is one whose merits have depended on whether and to what extent we envisaged the people concerned would be trustworthy as

277

well as adequately informed. You could make any of them
sound sinister by imagining a case where the motives of
the scientist were unworthy. Conversely, almost any can
be envisaged as a duty of compassion in certain defined
circumstances. This said, however, we find ourselves
forced to recognize, sadly, that in a fallen world legislation may have to be framed for, if not the worst case,
at least a far less ideal case, than if everyone were
guaranteed to have only the most transparent intentions
and the best of motives. In all our discussing and thinking we must be careful to distinguish between what might
be legitimate in God's sight - perhaps, in particular
cases, obligatory in God's sight - and what ought to be
made legal.

THE CHRISTIAN CHURCH

What then should the Christian church be doing? First,
the church might redeem its past by becoming the champion
of science in areas where fearful and less informed people
might perhaps oppose scientific research. It is essential,
however, for the church to be a critical champion. It
should criticize in love, and be merciless if there are any
signs of unbiblical tendencies. The implication would be
that the church should oppose research only if it infringes
biblical principles, or if the reasearch would take the
place of and prevent our doing something still better,
something more glorifying to God. This last point may be
important. There are always going to be enthusiastic people who are bitten with an idea and want to sell it. To
argue that there is nothing in the Bible against it is
not good enough. Part of our responsibility as Christians,
as indeed of anyone else in an effective community, is to
consider whether there isn't something still better, or
more urgent, that needs doing. We have to do our homework
before we can be clear that it would be still better (more
glorifying to God), but it is certainly part of our obligation to ask.

Secondly, a major responsibility of the church is to
clarify some key concepts in the debate. By the church
I mean Christian people. I don't necessarily mean parsons,
let alone general assemblies. But qualified Christians
ought to be busy, for example, working in what is at the
moment a live area in philosophy, seeking to clarify such
concepts of human nature, the person, human rights, consciousness, death. What rights can meaningfully be assigned to a fetus? Must a body which shows no signs of a
continuing conscious personality be preserved because it

is biologically alive? There is a huge package of concepts
that needs clarifying.

In another context there is a continuing need to clarify the concept of chance. Its innocent technical use in
science needs to be distinguished from that of its pagan
metaphysical namesake. To speak of the "Rule of Chance",
for instance, as if chance were an alternative agent to
God, can be grossly misleading, as well as scientifically
unjustified. What the scientist means by chance is simply
that which could not have been predicted on the basis of
prior data. So when geneticists speak of "taking a hand
from the genetic deck of cards" they must not be taken to
be advocating a pagan theology. The metaphysical overtones have no basis in their physical image of the process.
Moreover as far as the Bible is concerned, when "the lot
is cast into the lap, the whole disposing thereof is of the
Lord" (Prov. 16:33). In that sense, chance is a biblical
concept without any pagan overtones.

Then take the concept of liberty. Several papers
brought out the need for deeper analysis of this notion
in the present context. For the Christian, liberty does
not mean just doing one's own thing; it also means "being
subject to one another, for the sake of Christ" (Ephesians
5:21). This is a humbling yet richly rewarding concept of
liberty. Where the world in general thinks in terms only
of absence of restrictions, the church should have much to
contribute by way of a corrective emphasis. By the same
token, current uses of the concept of equality, penetrat-
ingly explored by Dr. Sinsheimer, need evaluation and il-
lumination in biblical terms.

Another task for Christians could be to promote and spell
out in detail, the implications of what David Allen called
the "principle of reciprocity". "Would I want the same
done to me?", he asked us. In sufficiently clear-cut cases
that is a good test. But of course there are awkward cases.
If we are considering whether a fetus with Down's syndrome
should be allowed to develop into a mongol child, there is
little help in asking "Would I like it done to me?" I
can never know what it would be like to be a fetus or a
mongol. There are a lot of borderline areas and gray areas
where the outworking of the principle of reciprocity is far
from clear.

Again, we are reminded by William Wilson that protecting
the right to treatment might be as important as protecting
the right to refuse treatment. Since the latter finds more

advocates at present, Christians might well be on the
alert to safeguard people's rights to the treatment that
could help them.

Dr. Perry London gave us a text on which perhaps the
church might well preach from time to time: "Only the
responsibility for the future of man rests with man: not
the future of man". We are responsible for what we can
do to shape our future; we are not responsible for the
real future. Responsibility for our future rests with God.
This might be an interesting sermon to preach, because
the distinction is not often observed in either utopian or
anti-utopian literature.

Finally, Carl Henry suggested that one of our prime
functions as Christians is to seek to "sensitize the conscience of the nation". No evangelical with a sense of
history could dissent from this. At the same time we would
do well to be wary here of the subtle and seductive temptations of scaremongering. There are many in our day who
make a reputation out of being scaremongers. Their books
sell because of the shivers they send down people's spines.
Works of this kind, when they obscure the factual issues
in clouds of emotional fog, bring despair to those fighting for proper and intelligent safeguards against the abuse
of science. Christians must beware of jumping on the bandwagon of the scaremongers. It is a temptation, perhaps
especially to evangelicals who have awakened suddenly
to their social responsibilities, to be mere echoes of
contemporary doomsmen , rather than critics of the critics.
Most critics today use essentially pagan criteria. Christians do not help by uncritically echoing them.

In this respect the church has surely its part to play
in the most difficult part of this whole enterprise for
our society, namely, learning what to want. The theory
of behavioral manipulation makes it clear that the greatest
power lies in the hands of the man who can determine what
we want. This is, therefore, a sensitive and fateful area
of the discussion on human engineering. What ought we to
want? It is important for the Christian not to take the
stance of the man who knows what he wants, and other people have just to listen. We will have to be ready to listen
just as much as the non-Christian, even though our ear is
bent primarily in the direction of God's word.

One more note of warning. The church needs to be wary
of ganging up with groups who do not respect God's priorities and pursue them with all their hearts. We can quickly

find ourselves trapped in unrealistic compromise. We may then be rightly stigmatized as letting the group down if at some later point it becomes clear that it does make a difference whether or not you believe that man's chief end is to glorify God and to enjoy Him forever. Equally, we must pay specially loving attention to any misgivings expressed by those in the church who are not equally informed, and may be more hesitant and fearful than we. The function of the church as salt in the earth is a corporate one. Our thinking in this area must be a fully corporate enterprise if it is to be fully open to such guidance as the Spirit of God can give His church. The more conservative and fearful are equally members of His body; whatever their difficulty in becoming articulate in our terms, we have no right to expect that His Spirit is going to be given more to us than to them in seeking the path of wisdom for His church.

I have tried in these reflections to indicate how the balance has swung, first one way and then the other, during our deliberations. Above all, what I heard us say to one another was: Let us be positive. This I think is not trivial. It was not at all to have been assumed in advance that a gathering of predominantly evangelical Christians should have consistently sought for positive good to come out of these new developments, one after another, and to have acknowledged by implication our obligation to further this positive good as God would enable us. It is remarkable, I think, that we had so much agreement. I trust and pray that it augurs well for evangelical involvement, with all the humble fear and trembling that Paul commands, in the development of legitimate human engineering for the good of man and the glory of God.

FOOTNOTES

[1]Luke 12:18-20.

EVANGELICAL PERSPECTIVES ON HUMAN ENGINEERING

Report of the Commission[1] from the
International Conference on Human Engineering
And the Future of Man
July 21-24, 1975
Wheaton College
Wheaton, Ill.

RATIONALE

We hold that all things are originated and sustained by
the creative will and action of God, as the Scriptures teach.
Therefore the world is neither to be deified nor vilified,
but rather is to be accepted as a trust from God. Human
stewardship includes the search for understanding of the
world. The Christian may confidently engage in this search
because all truth is God's truth.

God charged man with governing and developing His creation. Scientific investigation and application assist in
fulfilling this Divine mandate. Jesus summed up human
responsibility in two commands: to love God with all one's
being, and to love one's neighbor as one's self. Consequently, Christians should strive with compassion to ameliorate
the host of evils and suffering which entered the world
through the Fall.

Among the many accomplishments of science certain techniques of genetic, neurological, pharmacological and
psychological modification of human beings have great potential to enhance or erode their capacity to fully love God,
neighbor and self as directed by the Scriptures. Because
human beings are both finite and sinful special care must
be taken to use the tools of these technologies with humility and integrity, and to preserve and foster the freedom,
dignity and spiritual responsibility of man, who is created
in the image of God. Moreover, God calls for justice to be
reflected by the laws of the state in its treatment of
those potentially affected by this technology, as well as

283

toward and by those involved in developing and applying the technology.

Because no one is reducible solely to his biological, psychological or behavioral dimensions, or any combinations thereof, the results to be expected from human engineering will be limited. While God may benefit humanity through such technological processes, salvation is effected only by the supernatural grace of God through Jesus Christ. Nevertheless, human technology ought to be used responsibly by Christians in carrying out their concern for the needs of humanity.

PRINCIPLES

Much has been said in the past few years regarding the ethics of research and application of these technologies. In general, the techniques that were discussed at the conference, if appropriately applied, can be used to increase or restore man's physical and mental health. On the other hand, if used to excess, misapplied, or used in a careless or inconsiderate way, they can be harmful. In light of a Christian understanding of personhood, steps must be taken to insure respect for human rights and responsibilities:

1. We should consider available investigative procedures and treatments and carefully identify those which are applicable to a given individual's condition. Procedures with more drastic or irreversible effects must be more extensively justified by showing that beneficial results could reasonably be expected to outweigh deleterious effects. The underlying aim must be to improve the person's wholeness. To illustrate, psychosurgery applied to the treatment of a childhood behavior disorder resulting from a chaotic home environment is almost always a misapplication. Its use on a brain-damaged eighteen-year old who impulsively and homicidally attacks any person who tries to direct his behavior might be appropriate.

2. Investigatory procedures, applications of technology and release of data potentially harmful to an individual should be subject to the informed consent of the person primarily affected. In cases of those lacking capacity for informed consent to be exercised, special safeguards should be provided to protect their rights and interests.

3. Since all persons are made in the image of God we should treat everyone's claim with equity. There should

284

be no prejudicial consideration of any factor that sets one person apart from another. Christians have a special obligation toward those who are without power, and should insist on a fair access to beneficial technology for everyone.

4. The biblical principles of love and justice, among others, are essential bases for decision-making in this area. However, many choices regarding the application of these technologies do not follow easily from these guidelines. Often there are a variety of ethical concerns which bear on a single decision and pull us in diverse directions. Special concern should be given to conflicts between individual and social consequences of a particular decision. For example, genetic planning should consider, among other things, the implication of bearing and not bearing children, the risks of bearing children who may be seriously deformed, and the social consequences of caring for deformed children.

RECOMMENDATIONS

Given the complexity and urgency of these issues, we urge governmental bodies, scientific and professional groups and the general public to accelerate their efforts to exchange information and concerns about these matters. Reports such as those prepared by the Subcommittee on Health of the Committee on Labor and Public Welfare of the United States Senate (May 1975), the National Academy of Sciences (1975), Hastings Institute, and other scientific, professional, legislative and religious bodies ought to be widely distributed to individuals and groups from a variety of training and orientations. We particularly urge those who are followers of Christ to become deeply involved in this dialogue and subsequent decision-making.

Within the scientific community much benefit would come from more extensive discussions of ethical matters. We, therefore, encourage the following steps:
1. The training of scientists in the ethical dimensions of research and application;
2. The increasing consideration of ethical issues in scholarly journals;
3. The development of conferences and books that would help sensitize the professionals to ethical implications and decision-making approaches;
4. The rapid and vigorous implementation of plans to sponsor conferences and projects on ethical aspects of human engineering technology by such funding agencies as the National Science Foundation's Ethical and Human Value

Implications of Technology program. Such funding should
include groups with relatively diverse and relatively
homogeneous value orientations;

5. The provision of funds for research projects which
deal directly with ethical dilemmas, and which reach across
disciplinary boundaries. A current need for such study
exists with the research on recombinant DNA molecules, with
its many implications for human gene therapy.

Within the general public unnecessary alarm and ignorance
of developments must be avoided. From the local level to
the national arena, opportunity must increasingly be given
for citizen review of research design and the application
of human engineering techniques. These groups should look
beyond the immediate activities of researchers and practitioners to consider the future implications of present
inquiry. The legislative, executive, and judicial branches
of government increasingly will need to deal with bioethics.
Anticipating the implications of human engineering technology will, however, reduce the need for disrupting
research progress with governmental moratoria.

The Christian has extensive opportunities to participate
in public dialogue, employing the sensitivity of Christian
values and the leading of the Holy Spirit. We, therefore,
urge Christians in the spirit of love and service to make
themselves available for review boards and dialogue bodies
at all levels. Christian organizations, particularly the
professional and academic societies, are urged to deal with
the ethical aspects of human engineering in conferences
and publications. Finally, we urge Christian colleges and
institutions to develop their special calling to integrate
Christian ethics with scientific concerns.

FOOTNOTES

[1] Commission members were:

David F. Allen, M.D.
Yale University
New Haven, CT

Rodger K. Bufford, Ph.D.
Psychological Studies Institute
Atlanta, GA

V. Elving Anderson, Ph.D.
University of Minnesota
Minneapolis, MN

Paul W. Clement, Ph.D.
Fuller Graduate School of
Psychology
Pasadena, CA

Peter DeVos, Ph.D.
Calvin College
Grand Rapids, MI

Craig W. Ellison, Ph.D.
Westmont College
Santa Barbara, CA

Millard Erickson, Ph.D.
Bethel Theological Seminary
St. Paul, MN

Paul D. Feinberg, Ph.D.
Trinity Evangelical
Divinity School
Deerfield, IL

Lon Fendell, Ph.D.
Legislative Assistant
Hon. Mark Hatfield, United
States Senate

Norman L. Geisler, Ph.D.
Trinity Evangelical
Divinity School
Deerfield, IL

Carl F.H. Henry, Ph.D.
World Vision International
Pasadena, CA

Paul B. Henry, Ph.D.
Calvin College
Grand Rapids, MI

Robert L. Herrmann
Oral Roberts University
School of Medicine
Tulsa, OK

Donald M. MacKay, Ph.D.
University of Keele
Staffordshire, England

James H. Olthuis, Ph.D.
Institute for Christian
Studies
Toronto, Canada

John A. Olthuis, Ll.B.
The Committee for Justice
and Liberty Foundation
Toronto, Canada

John Scanzoni, Ph.D.
Indiana University
Bloomington, IN

Richard L. Spencer, Ph.D.
Fuller Theological Seminary
Pasadena, CA

William P. Wilson, M.D.
Duke University Medical
School
Durham, NC

Michael J. Woodruff, Jur.D.
West and Favor
Attorneys-at-Law
Santa Barbara, CA

ORGANIZATIONS COSPONSORING INTERNATIONAL CONFERENCE
ON HUMAN ENGINEERING AND THE FUTURE OF MAN

THE AMERICAN SCIENTIFIC AFFILIATION

The American Scientific Affiliation (ASA) is a fellowship of men and women of science who share a common fidelity to the Word of God and to the Christian faith. It has grown from a handful in 1941 to a membership of over 2,000 in 1977. The stated purposes of the ASA are "to investigate any area relating Christian faith and science" and "to make known the results of such investigations for comment and criticism by the Christian community and by the scientific community."

For over thirty years, the ASA has worked at the significant task of integrating scientific studies of the natural world with God's special revelation of Himself through the Bible. ASA members believe that honest and open study of God's dual revelation, in nature and in the Bible, eventually lead to understanding of its inherent harmony.

The purposes of the organization are carried out primarily through publications and meetings. The affiliation has two regular publications; the Quarterly JOURNAL OF THE ASA and the bimonthly NEWSLETTER, as well as occasional monographs and reprints. There is a national annual meeting held each August in a different part of the country and eighteen organized groups hold meetings throughout the year.

For further information, write to: 5 Douglas Ave., Elgin, Illinois 60120.

THE CHRISTIAN ASSOCIATION FOR PSYCHOLOGICAL STUDIES

The Christian Association for Psychological Studies (CAPS) was formally organized in 1955 with several Christians working in psychology and related fields. They had been meeting informally for two years for fellowship and discussion of mutual concerns. CAPS has grown to an international organization of approximately 1000 members, with regional chapters throughout the United States. It is incorporated under the laws of the State of Michigan as a non-profit organization.

The Christian Association for Psychological Studies is comprised of profesionals in psychiatry, medicine, the ministry, sociology, social work, education, guidance, and related disciplines. CAPS aims to help those professionals cooperate in the search for a better understanding interpersonal relationships, and to integrate a Christocentric frame of reference within these disciplines. CAPS holds an Annual Convention, publishes a quarterly BULLETIN and publishes a Membership Referral Directory. In addition it has begun joint publication of a Christian Perspectives on Counseling and the Behavioral Sciences series with Harper and Row. The series attempts to describe and analyze relationships between Christianity and different aspects of psychology.

For further information write to: 26711 Farmington Road, Farmington Hills, Michigan 48018.

THE CENTER FOR THE STUDY OF THE FUTURE

The Center for the Study of the Future was established in 1973 by Carl Townsend, and is based in Portland, Oregon at 4110 N.E. Alameda.

The Center believes that God is working in history redeeming the world and man to himself. God is acting to heal, to change, and to bring man to full actualization. The Center believes that to be aware of this pattern of acting is the first step in becoming fully human.

GOALS
1. Resource and educational center for Christians interested in studying the future and how the Body of Christ should relate to it.
2. Research center for Christians wishing to experiment with risk concepts of mission, worship, and koinonia.
3. Celebration center for those wishing to celebrate change and its relationship to the working of Christ in their lives.
4. Healing ministry for persons experiencing difficulties in relating to patterns of change that are moving in their lives.

The ministry is open to all faiths. It is chartered as a non-profit organization in Oregon.

Current ministries include a monthly newsletter, PATTERNS, which is written for change agents in today's church and is available on a subscription basis for $10/year. The Center

290

is also developing an information retrieval system using a microcomputer that can be purchased quite inexpensively for churches and service organizations. A computerized bibliography of articles dealing with human engineering, values and the future is available through CSF.

THE CHRISTIAN COLLEGE CONSORTIUM

The Christian College Consortium was established in 1971 to provide a medium for the development and implementation on a national basis of cooperative programs. All Consortium programs are designed to reinforce the unique purposes of the member institutions, with primary consideration always given to the implications and imperative of the Christian worldview in higher education.

Consortium members are committed to the tenets and spirit of evangelical Christianity; are in earnest about the integration of Christian commitments in all areas of activity. All are four-year liberal arts colleges with full regional accreditation. To be fully Christian today in scholarship, learning, and teaching demands an investment of time, energy and resources by the member institutions.

Consortium actitivies are planned to increase learning opportunities for students, encourage Christian educational development for faculty and staff members, improve administrative efficiency, and advance the cause of evangelical higher education in the nation. The member colleges take seriously their responsibilities to provide needed alternatives in American highter education. The Consortium is supported by member college dues and fees, by foundations and corporations, and by individual donors.

Further information can be obtained from 1755 Massachusetts Ave., N.W., Washington, D.C. 20036.

THE CHRISTIAN MEDICAL SOCIETY

The Christian Medical Society is a professional society of almost 4,000 physicians, dentists, medical and dental students who have joined together in order to enhance their commitment to Christ both within and through the medical and dental professions. Primary concerns of the society include sharing the hope of a Christian faith with colleagues and patients, serving humanity either at home or abroad with a professional excellence that is coupled with God's loving concern, integrating the various contemporary issues in biomedical ethics with historic Christian ethics, and providing a forum where members and friends of the society may explore all of these opportunities in an atmosphere of mutual professional respect and personal Christian commitment.

There are about 185 chapters of the society located for the most part near medical complexes in the United States. Members in good standing of the medical and dental professions are eligible for membership as they are able also to express their profession of faith. All members receive CMS publications which include the bimonthly JOURNAL and the bimonthly NEWS AND REPORTS. The society is governed by a member-elected House of Delegates which is convened each year in May; salaries and expenses are financed through the gifts and dues of the membership.

Occasional symposia have addressed themselves to the ethics of abortion, birth control, sterilization, demon possession and the occult. Cooperative ventures such as this conference on human engineering have been a part of CMS history; international conferences with similar organizations around the world have been held triennially: Amsterdam (1963), Oxford (1966), Oslo (1969), Toronto (1972), and Singapore (1975).

For further information write: 1122 Westgate (P.O. Box 890), Oak Park, Illinois 60301.

THE EVANGELICAL THEOLOGICAL SOCIETY

The Evangelical Theological Society was organized in 1949 by some sixty Bible-believing scholars in North America. These scholars organized annual meetings for biblical and theological discussion. While denominational loyalties and doctrinal orientations are widely divergent, there has been no disposition whatever to compromise on the one matter which all members consider basic: the inerrancy of Scripture.

The membership has increased from sixty in 1949 to more than one thousand in 1977. Anyone who possess a Th.M. degree or its equivalent and who holds to the doctrinal basis of the Society (the Bible alone and the Bible in its entirety is the word of God written, and therefore inerrant in the autographs) may seek membership.

The Society publishes the JOURNAL OF THE EVANGELICAL THEOLOGICAL SOCIETY on a quarterly basis. The JOURNAL contains articles read at annual or regional meetings, book reviews and announcements. Besides, the Society publishes scholarly monographs.

By means of meetings and publications, the Society seeks
to fulfill its purpose: "...to foster conservative biblical
scholarship by providing a medium for the oral exchange and
written expression of thought and research in the general
field of the theological disciplines as centered in the
Scriptures."

The Society has sponsored two daughter organizations:
the Near East Archaeological Society and the Evangelical
Philosophical Society. These two organizations meet con-
currently with E.T.S.

THE INSTITUTE FOR ADVANCED CHRISTIAN STUDIES

The Institute for Advanced Christian Studies (IFACS)
was established to foster consultation, research and
writing by evangelical scholars. Underlying this purpose
are two convictions: 1) that the Christian world view needs
wider enunciation in the face of contemporary secularism;
2) that the Bible needs to be applied as the guide to faith
and practice on all fronts in today's problems.

Since its founding in 1966, research funds have been
awarded to 22 scholars and four conferences have been supported. From the resulting research and writing, a number
of books and journal articles have been published or are in
preparation. Academic disciplines represented include political science, sociology, linguistics, philosophy, biblical
studies, and theology.

The Institute was originally funded by Lilly Endowment,
Inc., but seeks to carry on its work through wide support
from interested Christians. It is not officially identified
with any denomination. Scholars are chosen without reference
to church affiliation. The organization maintains a mail
address in Chicago, but owns no real estate, pays no rent,
and has no employees. Its directors are scholars who serve
without salary. Contributions are recognized as tax-exempt.

Copies of a ten-year summary of the Institute's program
and finances are available by writing to Box 95496, Chicago,
Illinois 60690.

THE INSTITUTE FOR CHRISTIAN STUDIES

The Institute for Christian Studies, 229 College Street,
Toronto, Canada, is an interdisciplinary graduate center for
research and teaching. The Institute focuses on the examination of questions which underlie various academic disciplines

and serve as the points of communication among fields. These foundational matters pertain to the philosophy, the methodology, the general theory, and the history of an academic field of study. We are interested in investigating the terrain in which religion, philosophy, history and the special sciences interpenetrate. We seek to develop an integrated understanding of learning and of the relation of learning to life as a whole.

The nine members of our interdisciplinary-philosophical faculty lead seminars in the philosophy, theory, and history of these areas: Systematic Philosophy, History of Philosophy, Aesthetics, History and Historiography, Political Theory, Philosophical Theology, Systematic Theology, Psychology and Socio-Economics.

The Institute approaches academic study from a Christian perspective. We explore the implications for scholarship of Jesus Christ's liberating re-creation of the whole of life. In choosing to develop a Christian perspective in learning we pursue a line of thought alternative to most of the variety of perspectives common in today's secular universities.

The Institute offers four ways of organizing a course of study. The Master of Philosophy program; the Doctor of Philosophy program; the Certificate in Christian Studies program; and Non-program study. The Institute programs help students to develop their world view and prepare them for careers in teaching and scholarship, the pastorate, social and political work, journalism and writing, the arts, industry, counseling and other fields.